Springer Proceedings in Mathematics & Statistics

Volume 397

This book series features volumes composed of selected contributions from workshops and conferences in all areas of current research in mathematics and statistics, including data science, operations research and optimization. In addition to an overall evaluation of the interest, scientific quality, and timeliness of each proposal at the hands of the publisher, individual contributions are all refereed to the high quality standards of leading journals in the field. Thus, this series provides the research community with well-edited, authoritative reports on developments in the most exciting areas of mathematical and statistical research today.

Isadora Antoniano-Villalobos ·
Ruth Fuentes-García · Lizbeth Naranjo ·
Luis E. Nieto-Barajas · Silvia Ruiz-Velasco Acosta
Editors

Interdisciplinary Statistics in Mexico

AME Virtual Meeting, September 10–11, 2020, and 34 FNE, Acatlán, Mexico, September 22–24, 2021

Selected Contributions

Springer

Editors
Isadora Antoniano-Villalobos
Department of Environmental Sciences, Informatics and Statistics
Ca' Foscari University of Venice
Venice, Italy

Ruth Fuentes-García
Department of Mathematics
Facultad de Ciencias, UNAM
Mexico City, Mexico

Lizbeth Naranjo
Department of Mathematics
Facultad de Ciencias, UNAM
Mexico City, Mexico

Luis E. Nieto-Barajas
Department of Statistics
ITAM
Mexico City, Mexico

Silvia Ruiz-Velasco Acosta
Department of Probability and Statistics
IIMAS, UNAM
Mexico City, Mexico

ISSN 2194-1009 ISSN 2194-1017 (electronic)
Springer Proceedings in Mathematics & Statistics
ISBN 978-3-031-12777-9 ISBN 978-3-031-12778-6 (eBook)
https://doi.org/10.1007/978-3-031-12778-6

Mathematics Subject Classification: 62-06, 34F05, 60G15, 60J60, 60J75, 62B05, 62D05, 62F10, 62F15, 62F40, 62H11, 62H25, 62H30, 62J12, 62M05, 62N01, 62P05, 62P10, 62P15, 62P20, 62P25, 68T05, 91D20, 91D25, 92D30

This Springer imprint is published by the registered company Springer Nature Switzerland AG
The registered company address is: Gewerbestrasse 11, 6330 Cham, Switzerland

Preface

This volume includes a selection of peer-reviewed contributions presented at the 2020 Virtual Meeting of the Mexican Statistical Association (AME) which took place on September 10 to 11 and the 34FNE meeting held from 22 to 24 September, 2021, on the theme "Statistics in the multidisciplinary".

The FNE (Foro Nacional de Estadística) is the official meeting of the Mexican Statistical Association (AME), taking place annually since 1986. In 2018, the 33FNE was held together with the Joint Statistical Meeting of Latin America, the 13CLATSE (Congreso Latino Americano de Sociedades de Estadística). On such occasion, a collection of selected works was published as a book of proceedings entitled *Selected Contributions on Statistics and Data Science in Latin America.* The decision was made to hold the FNE biennially, publishing, each time, a selection of high-quality papers in a similar volume. The following edition was originally planned for September 2020 in Mexico City. But, as the date grew closer, amidst the Coronavirus disease (COVID-19) pandemic, it had to be postponed hoping the conditions would soon improve and allow an on-site event. The 2020 AME Virtual Meeting took place instead. One year later, with sanitary and travel restrictions still active throughout the globe, individuals and institutions sought new ways to adapt, to overcome the fear and paralysis, and to make efforts, however large or small, to remain active and connected. As many other events over the pandemic period, the 34FNE became a virtual meeting, and the Facultad de Estudios Superiores (FES) Acatlán, a decentralized faculty of the Universidad Nacional Autónoma de México (UNAM) took over the responsibility of organizing it.

The 2020 AME Virtual Meeting consisted of five plenary presentations, delivered by Massimiliano Caporin (University of Padua, Italy), David B. Dunson (Duke University, USA), Alejandro Jara (Pontificia Universidad Católica de Chile, Chile), Esther Ruiz Ortega (Universidad Carlos III de Madrid, Spain), and Yongcheol Shin (University of York, UK), and 12 parallel sessions on Inference, Networks and waiting times, Multivariate analysis, Risk management, GARCH models and risk, Biostatistics, Public health and safety, Statistical computing, Forecasting and the exchange market, Statistical modeling, and COVID and Education, for a total of 24 oral presentations.

The scientific program of the 34FNE included a total of 39 oral presentations, organized into seven contributed (COVID-19, Quick counts, Statistical learning and finance, Statistical learning and functional data, Finance, Social statistics, and Uncertainty Quantification and simulation), four invited (Statistics in the study of the COVID-19 pandemic, 2021 Quick counts, Paleo-statistics, and Census and national indicators), and six plenary sessions delivered by Roy Campos (Consulta Mitofsky, Mexico), Julia Palacios (Stanford University, USA), Arturo Erdely (FES-Acatlán, UNAM, Mexico), Gustavo Cano (Buró de Crédito, Mexico), Jorge Soberón (University of Kansas, USA), and Jesús Ramos (Sociedad Mexicana de Ciencia de Datos, Mexico). 20 poster presentations and 3 short courses on The use of the Python programming language for Statistics, held by Andrés Christen (CIMAT, Mexico), Gustavo Fuentes, and Alejandro Leonel Martínez (FES-Acatlán, UNAM, Mexico); Statistics and Epidemiology, held by Mario Santana (IMATE-Juriquilla, UNAM, Mexico); and The Julia programming language, held by Arturo Erdely (FES-Acatlán, UNAM, Mexico), completed the program.

For their great effort in difficult times, we thank Vladimir Rodíguez Caballero (ITAM) organizer of the AME 2020 Virtual Meeting, as well as all members of the FES-Acatlán local organizing committee for the 34FNE: Christian Delgado Elizondo, Mahil Herrera Maldonado (president), Luz María Lavín Alanís, and Omar Ulises Morales Dávila. The high quality of the 34FNE was ensured by the Scientific Committee: José Andrés Christen (CIMAT-Guanajuato), Mario Santana Cibrian (IMATE-Juriquilla, UNAM), and María González Videgaray (FES-Acatlán, UNAM). A special recognition is due to all participants and to the Mexican statistical community. The ongoing quality of their work gives substance to this volume.

It would not be honest to deny that scientific production was affected by the pandemic, and the work of the Mexican statistical community was no exception. But the work did not stop, the quality did not flounder. In a world full of uncertainty, especially for younger researchers, the mission of the FNE, to offer opportunities for statisticians and practitioners to share the latest developments in research and applications and to expand their collaborative networks within the country and internationally, is infused with renewed importance. We are confident that the developments presented in this volume will be of interest to academics and practitioners around the world.

Venice, Italy — Isadora Antoniano-Villalobos
Mexico City, Mexico — Ruth Fuentes-García
Mexico City, Mexico — Lizbeth Naranjo
Mexico City, Mexico — Luis E. Nieto-Barajas
Mexico City, Mexico — Silvia Ruiz-Velasco Acosta
June 2022

Contents

A Methodological Proposal to Model and Evaluate the Complexity of the Mexican Geo-Electoral System

M. David Alvarez-Hernández and Shaní Alvarez-Hernández

1 Introduction

To understand Mexico's geo-electoral system, it is necessary to first explain how it is defined. The concept of geo-electoral system refers to two main issues: (1) the political-electoral and socio-demographic characteristics of each territorial unit of the country and (2) the existing interrelationships between these territorial units (see [1]). In particular, electoral sections can be considered as the fundamental units of the Mexican geo-electoral system, since they represent the base entity from the administrative and political-electoral point of view.

In this context, an elemental and basic part of any geo-electoral system are the voter lists, since they are essential instruments to provide certainty and legitimacy to the political-electoral processes that take place in a country. The task of constructing these instruments acquires different degrees of complexity depending on the specific characteristics of the geo-electoral systems, since factors such as electoral regulations, socio-political conditions, and the particular socio-demographic and geographic characteristics of each nation come into play. In general terms, a voter list constitutes a list of citizens that is integrated from a series of conditions imposed by the electoral regulatory framework that governs its conformation, particularly with regard to (1) requirements that citizens must meet to be part of the list, (2) means of identification of the members of the list, (3) physical means of safeguarding the information of the list, and (4) mechanisms for updating and consolidating the information of the list (see [2]).

M. D. Alvarez-Hernández (✉)
Instituto Nacional Electoral, Dirección de Estadística, Mexico City, Mexico
e-mail: miguel.alvarez@ine.mx

S. Alvarez-Hernández
Justice Program, UNODC Mexico, Mexico City, Mexico

I. Antoniano-Villalobos et al. (eds.), *Interdisciplinary Statistics in Mexico*, Springer Proceedings in Mathematics & Statistics 397,
https://doi.org/10.1007/978-3-031-12778-6_1

In the specific context of the Mexican geo-electoral system, its list of voters is represented by the Federal Voters' Registry (*Registro Federal de Electores* or RFE), which is made up of two interrelated instruments: the Electoral Roll (Padron Electoral or PE) and the Nominal List of Voters (*Lista Nominal de Electores* or LNE). The PE represents the registry of all those citizens who at some point in time processed and received a Voter Registration Card (CPV). In contrast, the LNE represents the registry of all those citizens who have a valid voting credential and, therefore, are able to vote. Both instruments are permanent but are subject to purging and updating procedures.

The RFE has contributed to provide greater certainty to the political-electoral processes that take place in Mexico and has also contributed to provide greater legitimacy to the results of elections, since one of the main purposes of the RFE is to guarantee free access and exercise of political rights [2]. It is important to consider that the RFE is administered exclusively by the National Electoral Institute (*Instituto Nacional Electoral* or INE), the authority that prepares and supervises the registration procedures (Integration), credentialing (Identification), safeguarding of databases (Maintenance), and consolidation of information (Updating). In the following, and to simplify the writing, the acronym IIMU will be associated to the previous group of four tasks.

Considering this background, it is essential to have analytical tools that allow planning, directing, and evaluating the programs and institutional actions focused on the IIMU tasks, and that also take into consideration the geographic diversity of the country and the plurality of the socio-demographic characteristics of the population. Particularly, it is crucial to have indicators that synthesize and quantify the factors that affect the electoral complexity of the electoral sections. As will be explained in greater detail in the following section, the concept of electoral complexity is defined as the set of geospatial, demographic, and structural factors that hinder the work of integration, identification, maintenance, and updating (IIMU tasks) of the RFE.

Therefore, this paper presents a methodology of analysis that consists of elaborating, from a diverse group of geospatial, demographic, and structural indicators, a set of Electoral Complexity Indices (ECI) that allow to typify the electoral sections according to their degree of electoral complexity.

2 Methodological Framework

The topic of complexity has had a significant impact on the way scientific knowledge is conceived, which has generated some controversies, since in the opinion of some experts (see [3, 4]), the studies on complexity represent the possible beginning of a scientific revolution; in contrast, complexity represents for others a paradigm without solid scientific bases that justify their inquiries [5]. This diversity of opinions has enriched the field of study of complex systems, but it has also caused the emergence of multiple definitions, theories, and lines of research on complexity, without any achieving a consensus in the scientific community [6, 7].

Regarding the specific role of complexity in the social sciences, it is important to mention that the quantitative branches of the social sciences have already devised numerous methods and indicators that allow the quantification and characterization of social phenomena, to such an extent that in the notion of the functionalist paradigm, put forward and described by Burrell and Morgan [8], it can be established that systems of socio-economic indicators (metrics, indices, rankings, classifications, etc.) are a fundamental basis for generating solid descriptions of social facts, since the functionalist paradigm proposes that human behavior can be studied in a real world in which relationships are concrete and tangible, and therefore, it is possible to build objective and value-free social sciences, in which the social researcher distances himself from the object of study and rigorously analyzes it through the scientific method.

Therefore, socio-economic indicator systems are built with the goal of providing knowledge that is close to objectivity, opening the possibility of generating changes in social reality through the formulation of normative policies based on the information provided by the indicators. However, it is important to recognize that there are other approaches to the social that do not seek planned change or objective knowledge, nor do they renounce other paradigmatic options that seek to achieve a holistic or nominalist perspective of social phenomena [8].

On the other hand, the traditional way of constructing socio-economic indicators has been strongly influenced by the reductionist doctrine, so that many quantification methods in social sciences tend to generate aggregate measures of phenomena in order to simplify the analysis of these phenomena [9]. However, in a complex and structurally dynamic social system, any "average" description of phenomena is inherently ineffective in understanding and controlling the dynamics of these systems. This limits the degree of validity of socio-economic indicators as survey instruments or as tools for normative policy formulation and evaluation, as complexity cannot be erased without depriving the final statistical products of their information content and usefulness for both social scientists and policymakers [10].

2.1 *Conceptualization of Electoral Complexity*

Although the concept of complexity has been adapted and wielded in various fields of social sciences, its use in electoral issues has been scarce or with a predominantly computational approach. For example, Faliszewski [11] performs an analysis of the complexity of electoral systems from a computational perspective, emphasizing how the different rules of electoral competition and the voting preferences of the population influence the complexity to compute and determine winners in electoral contests. Our research does not emphasize the aspect of electoral competition and participation in the geo-electoral system, since the objective of the research is to focus on quantifying the complexity associated with the factors that directly affect the institutional actions aimed at maintaining and consolidating the RFE.[1]

As noted above, the RFE is made up of two lists: the Electoral Roll (PE) and the Nominal List of Voters (LNE). To integrate, identify, maintain, and update these instruments (IIMU tasks), it is necessary to constantly carry out processes of personal data collection, database purging, implementation of field studies to update the electoral cartography, and elaboration of analyses on demographic dynamics [1]. On the other hand, to achieve an adequate execution of the IIMU tasks of the RFE, it is necessary to consider and weigh—during the planning and budgeting processes—all those factors that complicate or obstruct the performance of such tasks. In other words, the complexity associated with the geospatial, demographic, and electoral characteristics of each of the base territorial units of the Mexican geo-electoral system must be analyzed and quantified [1].

Consequently, the concept of electoral complexity can be approximated as the set of geospatial, demographic, and electoral factors that directly impact the development and execution of the IIMU tasks of the RFE.[2]

It is important to mention that the emphasis on geospatial, demographic, and electoral factors is a product of data limitations, since the primary objective of this research is to provide a heuristic definition of the concept of electoral complexity.

Considering the approximation of electoral complexity proposed, the problem of how to quantify the concept for each of the base components of the Mexican geo-electoral system arises. To solve this problem, we first identify the base indicators that describe the dynamics of the factors that determine electoral complexity, since these will be the main information inputs to build a stratification of complexity at the sectional scale. In particular, three groups of indicators that INE constantly monitors for the planning and execution of the IIMU tasks of the RFE are analyzed: geospatial, demographic, and electoral.

Geospatial indicators describe factors such as the size and geographic dispersion of the population registered in the RFE. These factors are considered in the work of integrating and updating the RFE, as they provide crucial information for carrying out institutional tasks related to electoral processes (e.g., operations for the installation of polling stations, updating electoral cartography, surveys, and consultations).

Demographic indicators are associated with demographic dynamics factors; specifically, with the demographic changes noted in the LNE. The analysis of demographic dynamics is a fundamental element to determine the degree of electoral complexity of the sections, especially considering that the programs of identification, updating, and maintenance of the RFE (for example, the credentialing campaigns, or the identification of anomalous migratory flows) must be correctly coupled to the different demographic dynamics observed in the geo-electoral system.

---

[1] Another interesting research work, partially related to the topic of electoral complexity, is the one developed by Haynes [12], in which he introduces the theory of complex systems and its possible applications in public services management.

[2] In [1], a first approach to this approach to electoral complexity is made.

Electoral indicators are fundamentally related to the maintenance and updating activities of the RFE (for example, the level of correspondence between the LNE and the PE is an indicator that reflects the degree of updating of the RFE). It is important to note that the territorial units that present a deficient diagnosis in these indicators usually receive priority attention and a greater allocation of institutional resources, in order to solve the deficiencies detected in the indicators, and thus guarantee the consistency of the RFE for the conduct of the electoral processes [1]. Consequently, electoral indicators are fundamental to determine electoral complexity.

2.2 *Statistical Indicators for Quantifying Electoral Complexity*

For the quantification of Electoral Complexity, a set of six base indicators were considered that synthesize the essential geospatial, demographic, and electoral characteristics of each of the electoral sections.

1. Size of the electorate (LNE). This demographic indicator reflects the number of citizens who reside in the sections and who are able to exercise their vote. Therefore,
$$LNE^l = L^l \tag{1}$$
where L^l is the number of citizens registered in the LNE for the section l as of the December 2019 cutoff.
2. Electorate Density (Densidad_LNE). The population density of the electorate is a geospatial indicator that reports the number of citizens registered in the LNE per urbanized area, that is, this indicator provides information on the degree of geographic dispersion of the electorate. Estimates of the LNE and the surface area covered by the urbanization were used to obtain this indicator:
$$Densidad\_LNE^l = \frac{L^l}{A^l} \tag{2}$$
where A^l is the surface area (square kilometers) of the urban development of the section l as of December 2019.
3. Coefficient of Updating of the RFE (Razon_LNE_PE). This electoral indicator reflects the level of updating of the LNE and the PE, since it indicates the proportion of citizens who maintain a current registration to vote (i.e., it is the ratio between the number of citizens registered in the LNE and the number of citizens registered in the PE):
$$Razon\_LNE\_PE^l = \frac{L^l}{P^l} \tag{3}$$
with P^l equal to the number of citizens registered in the section's PE^l as of the December 2019 cutoff.

4. Annual Growth Rate of the LNE (TC_LNE_2019). This demographic indicator shows the population growth (or decrease) observed in the LNE of the sections. This indicator was calculated on an annualized basis:

$$TC\_LNE\_2019^l = \frac{L^l - L^l_{2018}}{L^l_{2018}} \tag{4}$$

with L^l_{2018} equal to the number of citizens enrolled in the LNE of the section l as of the December 2018 cutoff.
5. Coefficient of Variation of the LNE (Coef_Var_LNE). This demographic indicator, obtained from the analysis of the LNE time series,[3] provides information about the demographic variation observed in the sections:

$$Coef\_Var\_LNE^l = \frac{\sqrt{\frac{1}{N}\sum_{i=1}^{N}(L^l_i - \overline{L}^l)^2}}{\sum_{i=1}^{N}\frac{L^l_i}{N}} \tag{5}$$

with L^l_i equal to the number of citizens enrolled in the LNE of the section l as of the end of the month i of 2019.
6. LNE Monthly Autocorrelation (Autocorr_LNE). This electoral metric measures the degree of similarity between a time series and a past version of itself. For the purposes of quantifying electoral complexity, it is of particular interest to calculate the LNE autocorrelation with a monthly lag (i.e., the degree of similarity between the behavior of one month with respect to the previous one), since the elaboration of LNE forecasts (to determine, among other things, the potential demand for CPV) is ordinarily carried out monthly.[4]

$$Autocorr\_LNE^l = \frac{\frac{1}{N}\sum_{i=1}^{N-k}(L^l_i - \overline{L}^l)(L^l_{i+k} - \overline{L}^l)}{\frac{1}{N}\sum_{i=1}^{N}(L^l_i - \overline{L}^l)^2} \tag{6}$$

2.3 *Data Transformation, Electoral Complexity Indices, and Stratification*

The modern paradigm of the construction of socio-economic indicators is characterized by incorporating aspects of the holistic perspective, which consists of preserving the multifactorial and multidimensional nature of the systems with the purpose of

[3] In this indicator, the analysis of the time series was conducted from July 2019 to December 2019, since in that year there were electoral processes in the entities of Aguascalientes, Baja California, Durango, Puebla, Quintana Roo, and Tamaulipas, which implied that in those states IIMU activities were temporarily halted during the months prior to Election Day.

[4] In this indicator, the time series analysis was also conducted from July 2019 to December 2019, in order to not consider the months leading up to Election Day.

defining global quantification indicators that make minimal use of aggregation methods or subjective weighting [1].

Therefore, following the precepts of the modern paradigm for the elaboration of socio-economic indicators, a set of Electoral Complexity Indices (ECI) is constructed from the information contained in the six base indicators. These indices will make it possible to quantify the complexity of the geo-electoral system at the sectional level.

ECIs can be formally defined as “index numbers that assess the degree of electoral complexity present in a territorial unit of the geo-electoral system based on the synthesis (weighting) of the information contained in a diverse group of indicators” [1]. In other words, the ECIs quantify in a single numerical value the degree of electoral complexity associated with the behavior (represented in the six base indicators) of the geospatial, Demographic, and structural factors that determine electoral complexity [1].

Considering the above definition, it is derived that the ECIs are indicators that allow the establishment of order and magnitude relationships between the different territorial units of the geo-electoral system, since they make possible the elaboration of stratifications and ordered lists (rankings) according to the degree of electoral complexity. Likewise, the ECIs also make it possible to measure variations in electoral complexity caused by changes in the behavior of the base indicators.

For the elaboration of the ECIs, it is necessary to use a dimensionality reduction technique, since from the set of six base indicators it is desired to obtain a set of indices that weigh, in a non-subjective way, the influence of each indicator in the determination of electoral complexity.

Consequently, the PCA technique is used to construct the ECIs. This multidimensional data analysis technique is fairly standard in the fields of multivariate statistics and data mining and has been widely used in various study contexts involving the reduction or transformation of data to obtain synthetic indices [13–15].

The PCA provides an assessment that weighs the effect of geospatial, demographic, and electoral factors on the observed degree of electoral complexity. In other words, each Principal Component represents an ECI (i.e., a specific weighted sum of the base indicators), and the rating obtained by each section in the ECIs represents a degree of electoral complexity unique to that index. In mathematical notation, the degree of electoral complexity can be represented as follows:

$$ECI_i^l = \sum_{j=1}^{p} w_i^j x_j^l \tag{7}$$

where ECI_i^l is the degree of electoral complexity of section l, corresponding to electoral complexity index i, w_i^j is the weighting corresponding to the original variable j of the electoral complexity index i, x_j^l is the value of section l in the original variable j, while p is the number of original variables introduced in the PCA.

For the choice of the number of Principal Components extracted from the PCA, there are several criteria [16]. In particular, the present research uses the percentage criterion of the part of the variance explained by Principal Component to evaluate the synthesis capacity of the components. As will be shown in detail in the next section of the article, the results of the variance explained criterion allow us to justify the choice of the first Principal Component as the main ECI and, therefore, a national ranking of electoral complexity can be constructed from that index. Subsequent Principal Components (i.e., subsequent ECIs) are also used to obtain alternative assessments of electoral complexity, as each ECI brings a different weighting structure to the base indicators.

Despite the advantages offered by the PCA, it is important to carefully examine the results obtained with this method in order to assess the explanatory capacity of the ECIs and their interpretation in terms of the original variables. It will be shown below that for our case study, the results of the PCA method indicate that the Principal ECI alone cannot be used to generate a consistent electoral complexity ranking, since this index does not reach an adequate level of information synthesis for the base indicators (reflected in the percentage of variance explained by the first Principal Component), and therefore, it is not fully justified to use the Principal ECI alone as a metric for comparing the degree of electoral complexity presented by the sections.

In this context, the lack of statistical robustness of the electoral complexity ranking (based only on the main ECI) can be overcome by a stratification of electoral complexity at the section level, which makes simultaneous use of the information contained in a selected subset of ECIs (specifically the first three ECIs), in order to achieve a higher percentage of explained variance.

To generate the stratification, a K-Means algorithm was used, which is frequently employed in the field of unsupervised learning algorithms, since it provides a simple approach to approximate, for any number of strata, the partition that minimizes the total variance within the strata [17]. It is important to note that there are different stratification algorithms and methods. For example, an alternative that is frequently used to address univariate stratification problems is the Dalenius and Hodges method. In this paper, this method is not implemented because the results of the PCA indicate that the ECIs do not reach an acceptable level of information synthesis, so that any stratification based only on one of these indices will fail in the task of classifying the degree of complexity adequately.

On the other hand, to determine the optimal number of strata, it is necessary to perform a silhouette score analysis (also known as silhouette coefficient analysis), since it allows to analyze graphically (through the so-called silhouette diagram) the separation distance between the strata of a specific configuration [18].

In summary, the joint implementation of the K-means method and the analysis of the silhouette coefficients allows us to determine the optimal number and configuration of clusters to stratify the sections according to the ratings obtained in a subset of the ECIs. In other words, the clustering analysis allows for the construction of a stratification of electoral complexity at the sectional level.

3 ECI Construction and Stratification

3.1 Exploratory Analysis and PCA Implementation

Figure 1 presents the results of the exploratory analysis (panel of correlograms and histograms) carried out on the set of six indicators considered in the study at the sectional level.

The histograms in Fig. 1 (shown on the diagonal of the panel) show that the six base indicators—with the exception of the indicators Coefficient of Updating the RFE (Razon_LNE_PE) and Coefficient of Variation (Coef_Var_LNE)—present positive asymmetric distributions, which denotes that there is a significant number of electoral sections that have atypical geospatial, demographic, and electoral characteristics. In particular, the highly skewed behavior of the indicators Voter Magnitude (LNE) and

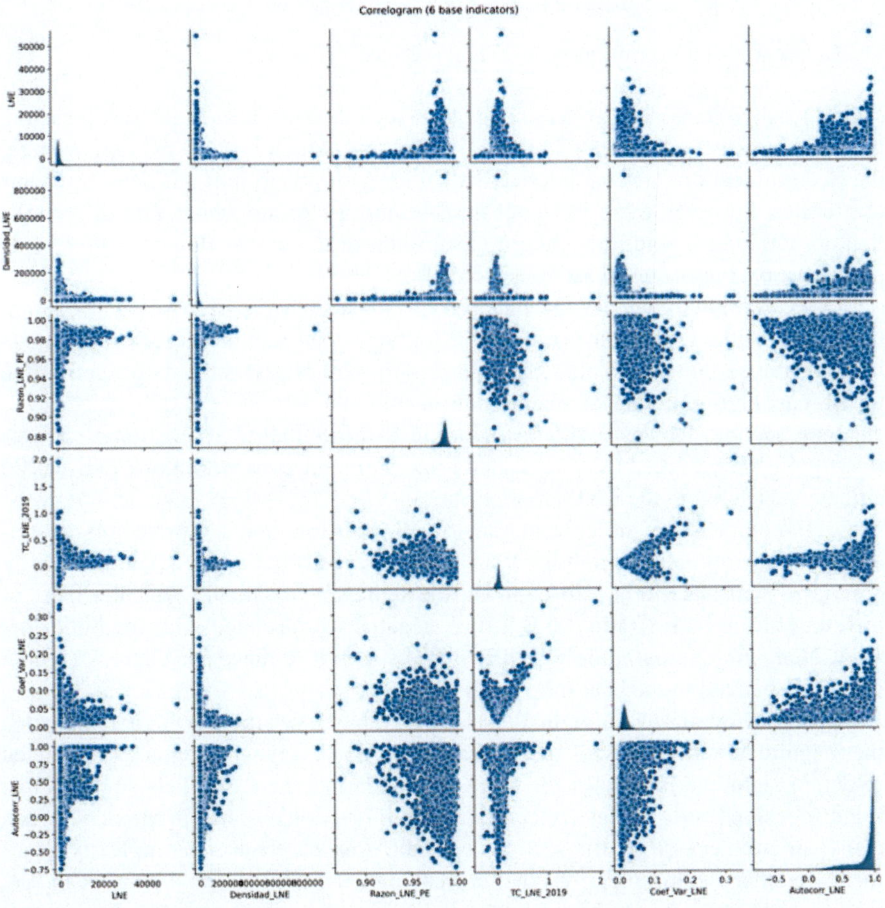

Fig. 1 Panel of correlation and distribution graphs of the indicators that form the sectional ECIs

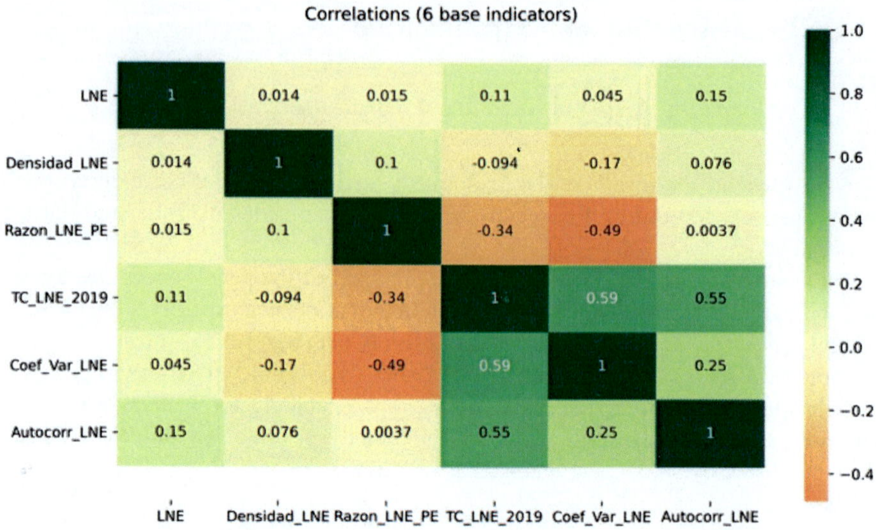

Fig. 2 Heat map of the correlation matrix of the sectional indicators

Voter Density (Densidad_LNE) stands out, since these two base indicators present a notorious positive asymmetry in their distributions, which implies that, on average, the electoral sections are characterized by having a low magnitude and density in their electorates; however, there is a considerable number of atypical electoral sections that do not comply with this pattern (each point in the scatter diagrams represents one of the 68,390 electoral sections studied).

As an example, consider the panel graph located in the first row and column (histogram of the LNE indicator). This histogram shows that most electoral sections have a reduced number of citizens registered in the LNE, and this is primarily due to the constitutional rule of maintaining a maximum limit of 3,000 citizens per electoral section. However, the histogram also shows that there are some sections that significantly exceed the established limit (some sections with more than 30,000 citizens registered in the LNE are identified).

On the other hand, the correlograms in Fig. 1 show that the indicators associated with demographic variability (TC_LNE_2019, Coef_Var_LNE) are positively correlated with each other. In contrast, the indicator associated with the level of updating of the RFE (Ratio_LNE_PE) is negatively correlated with the indicators associated with demographic variability. The strength of these correlations is more clearly visualized in the heat map in Fig. 2.

Figure 2 allows us to highlight that, at the sectional level, there is no indication that the magnitude and density of the electorate are positively—or negatively—related to demographic variability or the level of updating of the RFE. This suggests that highly densified and populated electoral sections (sections located in Mexico's urban areas) are not necessarily the sections with the greatest electoral complexity, since there are other electoral sections that present a greater demographic dynamism and, therefore, their degree of electoral complexity is also greater.

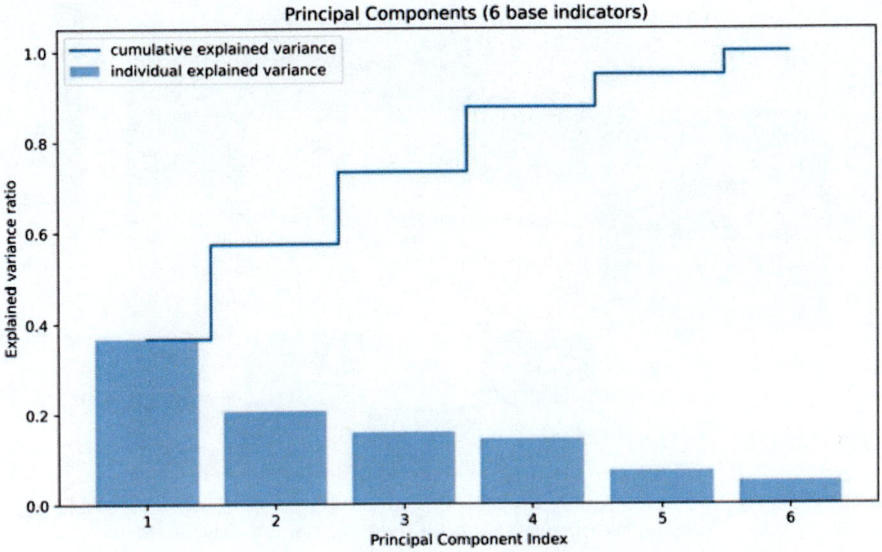

Fig. 3 Sectional ECI: Proportion of variance explained by principal component

Consequently, the results of the exploratory data analysis justify the relevance of applying the PCA method to the set of base indicators in order to obtain a new and more synthetic set of indices. The result of the percentage of variance explained by each Principal Component is shown in Fig. 3.

It can be observed in Fig. 3 that the first Principal Component (PC1) reaches a percentage of explained variance of 36%, and consequently, PC1 is taken as the main ECI. However, PC1 is not an indicator that fully synthesizes the characteristics of the six base indicators and, therefore, should be used with caution for the elaboration of a ranking that allows for comparisons of electoral complexity between electoral sections.

On the other hand, Fig. 3 also shows that the second Principal Component (PC2 or second ECI) contributes 20% to the percentage of variance explained, while the third Principal Component (PC3 or third ECI) contributes only 15%. Therefore, to make consistent comparisons of complexity between electoral sections, the construction of a stratification of electoral sections based on a subset of the Principal Components is justified, considering that, individually, each ECI does not achieve a satisfactory level of information synthesis.

In relation to the composition of the ECI, Fig. 4 shows the composition of the Principal Components to determine their interpretation.

For example, from Fig. 4, it can be seen that the base indicators that have the highest (positive) weight in PC1 (main ECI) are the Annual Growth Rate of the LNE and the Variation Coefficient of the LNE. Likewise, the Coefficient of Updating of the RFE also presents a significant (negative) weight within the composition of the main ECI.

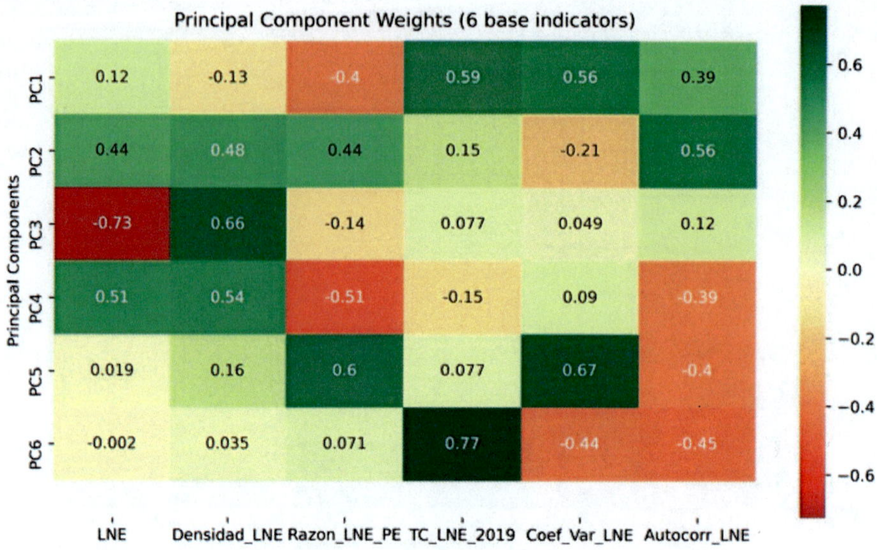

Fig. 4 Sectional ECI: Weights of the base indicators in each principal component

Considering that the base indicators that weigh most heavily in the composition of the main ECI are related to population growth and the temporal variations observed in the LNE, this index describes the electoral complexity associated with the demographic factor. In other words, the higher the (positive) valuation of an electoral section in the main ECI, the greater the temporal variability in its LNE, the higher the positive demographic growth, and the lower the level of updating between its LNE and the PE. On the other hand, the lower the (negative) valuation of an electoral section in the main ECI, the lower the variability of its LNE, the higher the level of updating of its LNE with respect to the PE, and the lower the demographic decrease in its LNE.

3.2 *Clustering Analysis and Stratification with K-Means*

From the results of the implementation of the PCA method, the K-means partition algorithm is used in the projection of the electoral sections in the space of the sectional ECIs, to perform a grouping analysis which allows to form groups of sections with similar degrees of electoral complexity and thus generate the stratification at the sectional scale.[5]

[5] The Euclidean metric is used as a measure of distance between the centroids of the clusters and the data points. Also, only the two-dimensional projection of the ECIs is plotted to facilitate the visualization of the silhouette plots.

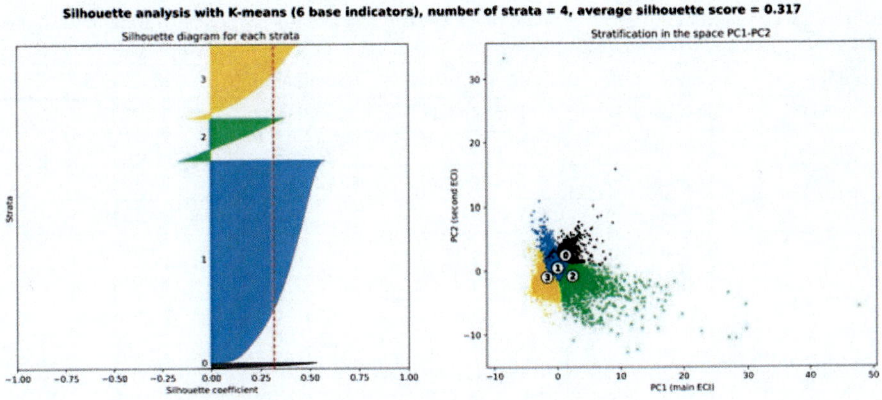

Fig. 5 Silhouette diagram for the 4-group configuration, and stratification in the subspace PC1-PC2

First, a silhouette coefficient analysis is performed to determine the optimal configuration of groups in the stratification. Figure 5 shows that the configuration of four groups can be considered optimal (compared to scenarios of 3 to 7 groups), since it reports the highest average silhouette coefficient (which in this case was 0.317). Details of the tests with different scenarios can be found in [19].

In the 4-group configuration, it can be observed that the stratification clearly distinguishes, along the direction of the PC1 axis, the electoral sections that have a positive evaluation in the main ECI (sections belonging to group 0 and 2) and the sections that obtain a negative evaluation (sections belonging to groups 1 and 3). Likewise, along the direction of the PC2 axis, the stratification separates the electoral sections that have a positive rating in the second ECI (groups 0 and 1) from the sections that have a negative rating (sections belonging to groups 2 and 3).

4 Electoral Complexity Ranking and Stratification Results

Given the number of electoral sections examined (68,390 sections), it is not possible to provide a complete summary of the results obtained in the total set of ECIs in this document. Therefore, Table 1 presents a sample of the sectional ranking of electoral complexity obtained from the results of the assessment of the sections in the main ECI. In particular, the first 10 sections with the highest level of electoral complexity are shown.

The information shown in Table 1 indicates that the electoral sections with the highest electoral complexity (with respect to the main ECI) are mainly located in geographic areas far from the main urban centers of Mexico (for example, six of the first ten electoral sections with the highest electoral complexity are located in the

Table 1 Sectional ranking of electoral complexity: Top 10 electoral sections by rating in the main ECI

Ranking	State	Municipality	Section	Main ECI
1	Mexico	Otzolotepec	6517	47.7232
2	Chihuahua	Uruachi	2696	29.7973
3	Sinaloa	Sinaloa	3427	29.7040
4	Chihuahua	Uruachi	2687	28.1519
5	Chihuahua	Bocoyna	147	27.0129
6	Nuevo León	Garcia	2547	26.9122
7	Chihuahua	Uruachi	2695	22.7558
8	Coahuila	San Juan de Sabinas	1711	19.7604
9	Nuevo León	General Escobedo	2737	18.4925
10	Quintana Roo	Tulum	948	17.7509

mountainous areas of Western Mexico) or in the limits of urban areas (as is the case of sections 2547 and 2737 in the state of Nuevo León).

On the other hand, Fig. 6 shows in detail the mapping of the sectional ranking of electoral complexity in the totality of the national territory. It can be observed that there are large areas of the national territory that present negative evaluations in the main ECI. This implies that in these geographic areas there are electoral sections that present little population variability in their LNE or that are characterized by negative population growth rates.

In contrast, there are other areas of the national territory that concentrate a significant number of electoral sections with a high positive valuation in the main ECI. This means that the electoral sections located in these areas show, in general, positive population growth rates, and they are also characterized by presenting a low level of updating in their LNE and PE.

With respect to the results extracted from the stratification, with the optimal configuration of groups (which receive the denominations G0, G1, G2, and G3), the electoral sections are grouped and the statistics of central tendency of each group are analyzed (by means of the violin diagrams shown in Fig. 7). This strategy allows us to identify the characteristic features of each stratification group and in turn allows us to identify and classify those electoral sections with an atypical degree of electoral complexity [20].

The violin plots in Fig. 7 show that the four stratification groups are characterized by similar statistics (median, interquartile range, and confidence interval) in each of the six base indicators, with the exception of the indicators Magnitude of Electorate (LNE) and Monthly Autocorrelation of the LNE (Autocorr_LNE), since in these indicators significant differences can be seen between the central tendency statistics of groups G0 and G3 and the central tendency statistics of the remaining groups.

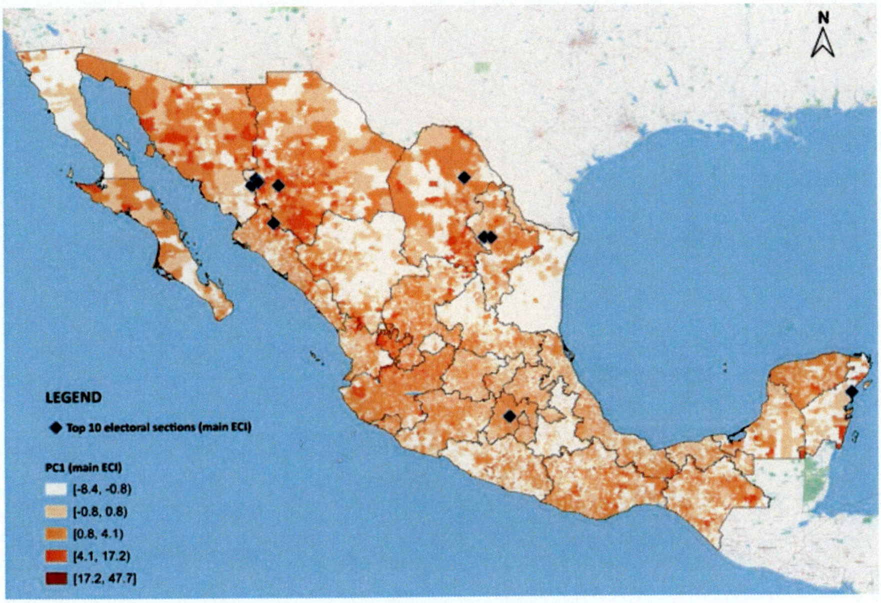

Fig. 6 Sectional ranking of electoral complexity

For example, the sections classified in group G0 have on average a larger electorate, but at the same time they present a population density similar to that observed in other stratification groups. On the other hand, the sections belonging to group G3 are characterized by a low level of monthly autocorrelation, although they do not present characteristics associated with demographic variability. And finally, some of the electoral sections classified in group G2 (see the elongated figure of the distribution) show a notorious demographic variability, as well as a low level of updating between the PE and the LNE.

On the other hand, the violin plots in Fig. 7 show that all the distributions of the base indicators are asymmetric, with different degrees and directions of skewness. In particular, the distributions observed in groups G0 and G2 exhibit the presence of long tails, as there are a high number of out-of-range polling sections (outliers), especially in the indicators related to the size of the electorate and to demographic variability (LNE, TC_LNE_2019, Coef_Var_LNE, and Autocorr_LNE).

It is important to highlight the atypical situation of some electoral sections, caused by the evaluation obtained in the sectional ECIs. These electoral sections are relevant to monitor since they present demographic and geospatial characteristics that significantly increase their level of electoral complexity. To identify those electoral sections that present an atypical valuation in the first three sectional ECIs, the interquartile range rule is used, and only the analysis of atypicality in the first three sectional ECIs is carried out, since these indices manage to explain approximately 72% of the

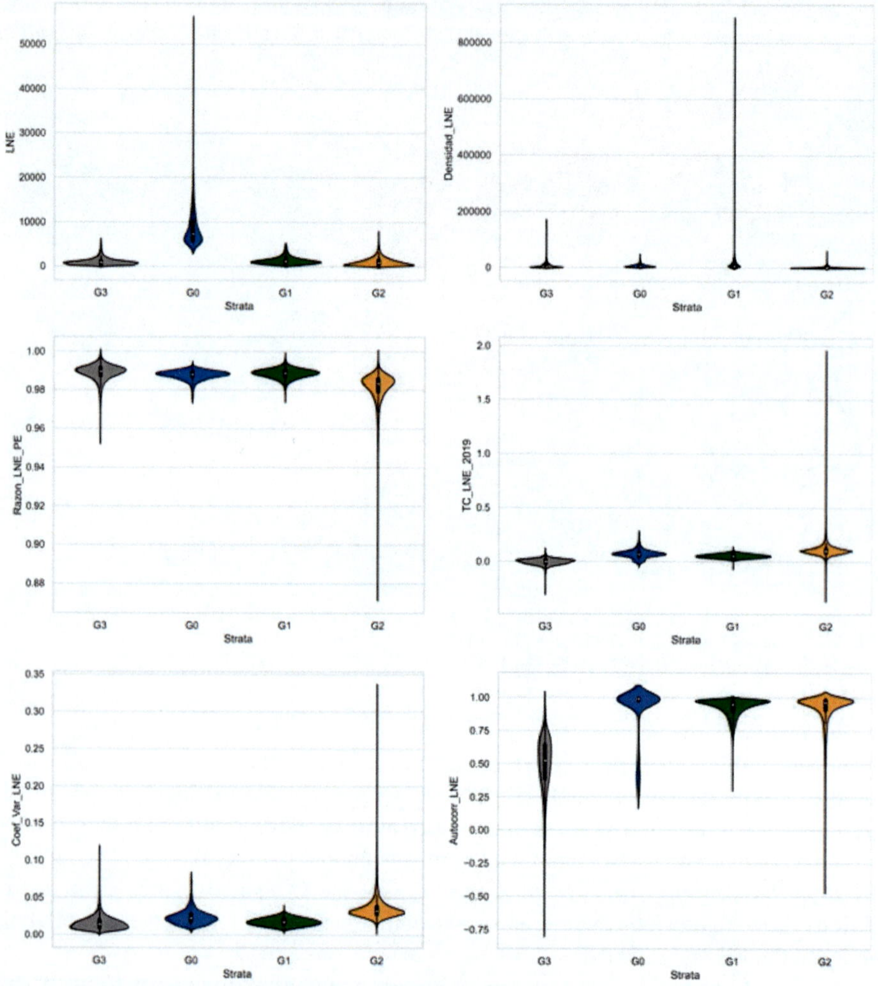

Fig. 7 Panel of violin plots of the six base indicators (with grouping according to stratification)

variance shown in the base indicators. In the attached files in [19], those electoral sections that show atypical electoral complexity are identified.

5 Conclusions

This research has allowed the design of a set of analytical tools that allow to quantify the complexity of the Mexican geo-electoral system at the sectional scale, with the objective of monitoring and evaluating more precisely the performance of the tasks of Integration, Identification, Maintenance, and Updating of the RFE. Likewise, from the holistic analysis of the base indicators, the results obtained provide a more reliable and accurate description of the current geospatial, demographic, and electoral characteristics present in the Mexican geo-electoral system.

These results were achieved thanks to a methodology based on three elements: the concept of electoral complexity, the set of base indicators selected to characterize the determinants of electoral complexity at the sectional level, and the multivariate analysis techniques for the construction of the ECIs and the stratification of electoral complexity.

The PCA method provided a robust mathematical procedure to handle the multidimensionality implicit in the set of base indicators, as well as to holistically model the information contained in the base indicators. In this way, the PCA made it possible to obtain a set of synthesizing indicators (ECIs) to combine and weigh each of the base indicators. However, the results obtained in this research allow us to conclude that the ECIs, considered individually, present some limitations to adequately synthesize the information contained in the base indicators.

Consequently, in order to establish better comparisons of electoral complexity between the different territorial units of the geo-electoral system, it is necessary to have a stratification that complements the complexity rankings obtained from the ECIs.

The first step to obtain the stratification of electoral complexity was to select the appropriate method to generate the stratification and to subsequently determine the optimal number of strata it should have. In this context, the analysis of the silhouette coefficients was used, obtaining results that allow us to affirm the following: the K-means grouping method is adequate to generate a consistent stratification of electoral complexity, being the optimal configuration of the stratification of four groups.

Based on the results generated by the ECIs and stratification, some relevant geographic patterns that emerge because of the different demographic and geospatial characteristics of the electoral sections were obtained and analyzed. For example, some groups of electoral sections of special interest for INE (atypical cases) were determined, since they present a notoriously higher degree of electoral complexity than the average of the sections, and therefore, it is necessary to monitor and attend to these sections in greater detail in order to guarantee the quality and functioning of the IIMU tasks.

In view of the above considerations, some recommendations and lines of future research are pointed out. Since the methodology for calculating the ECIs was subject to the volume of data examined and the available computational capacities, it is suggested that alternative methods of data transformation be used to advance in the analysis. In addition, an important line of future research would be to include other types of socio-economic and political-electoral indicators in the ECIs, since it is plausible to broaden the concept of electoral complexity to incorporate other factors that could potentially determine the complexity of the geo-electoral system. For example, indicators related to the demographic structure, migration flows (internal or external), voting results and preferences, or the degree of social development could be included. In this way, a more detailed analysis could be made of the social and political characteristics of the populations that are grouped into the different strata generated from the ECIs. Likewise, periodic updates of the ECIs and stratification are suggested (for example, at the beginning of the federal electoral processes) to continuously monitor the evolution of the base indicators and the effects of the institutional strategies aimed at improving the registration and updating services of the RFE.

This research ends with an observation. The topic of electoral complexity is far from being concluded, and therefore, the analysis of the complexity of the Mexican geo-electoral system can be considered a starting point for future studies on complexity in socio-economic systems, specifically on the topic of the management of public institutions and services from a complex systems perspective.

In this regard, it is important to remember that the theory of complex systems poses a fundamental challenge to the simplistic cause-and-effect thinking that predominates in the management practices of public administration [12]. One of the pillars of the complex systems paradigm is the notion that linear models are not adequate for analyzing complicated circumstances in which many variables interact over time, much less for determining causal relationships.

Therefore, decision-makers should ask themselves questions such as the following: Are large-scale statistical models really superior to individual, detailed case studies? What methods of data collection and interpretation are best suited to generate efficient management actions in the public administration? An even more general and far-reaching question could also be posed with respect to the evolution of the State's institutions: Does the current structure of public institutions adequately manage the complexity of socio-economic systems in such a way that the maximum level of social welfare can be achieved? Answering these questions will require a holistic methodology that takes the complex systems paradigm as its central point.

Notes and Comments. The results presented in this article were generated with the assistance of the machine learning models provided in the scikit-learn Python library. For a comprehensive explanation and demonstration of the implementation of the exploratory analysis and data transformation methods in scikit-learn, it is recommended to review [21]. Also, it is suggested to the reader to review [19] for a detailed analysis of the implementation of the code for the ECI generation and stratification.

References

1. Hernández, Á., David, M., Texocotitla, M.Á.: El sistema geoelectoral mexicano desde la perspectiva de los sistemas complejos. Apuntes Electorales 19(63), 213–258 (2020)
2. Welti Chanes, C.: Calidad del Padrón Electoral y elecciones confiables. Papeles de Población, 100, 13–52 (2019) https://doi.org/10.22185/24487147.2019.100.12
3. Waldrop, M.: Complexity: The Emerging Science at the Edge of Order and Chaos. Simon & Schuster, United States (1992)
4. Wolfram, S.: A New Kind of Science. Wolfram Media, United States (2002)
5. Horgan, J.: The End of Science: Facing the Limits of Knowledge in the Twilight of the Scientific Age. Addison-Wesley, New York (2015)

6. Becerra, G.: La Teoría de los Sistemas Complejos y la Teoría de los Sistemas Sociales en las controversias de la complejidad. Convergencia 27, 1–23 (2020). https://doi.org/10.29101/crcs.v27i83.12148
7. Byrne, D.: Complexity Theory and the Social Sciences: An Introduction. Routledge, London (2001)
8. Burrell, G., Morgan, G.: Sociological Paradigms and Organizational Analysis. Routledge, New York (2019)
9. Haynes, P.: Social Synthesis: Finding Dynamic Patterns in Complex Social Systems. Routledge, New York (2018)
10. Maggino, F.: Complexity in Society: From Indicators Construction to Their Synthesis. Springer, Switzerland (2017)
11. Faliszewski, P., Hemaspaandra, E., Hemaspaandra, L.A., Rothe, J.: A richer understanding of the complexity of election systems. Arvix (2006). https://arxiv.org/abs/cs/0609112
12. Haynes, P.: Managing Complexity in the Public Services. Open University Press, London (2003)
13. Abdi, H., Williams, L.J.: Principal component analysis. WIREs Comp. Stat. 2, 433–459 (2010). https://doi.org/10.1002/wics.101
14. Giuliani, A., Colafranceschi, M., Webber, C.L., Zbilut, J.P.: A complexity score derived from principal components analysis of nonlinear order measures. Phys. A: Stat. Mech. Appl. 301(1–4), 567–588 (2001). https://doi.org/10.1016/S0378-4371(01)00427-7
15. Jackson, J.E.: A User's Guide to Principal Components. Wiley, New Jersey (1991)
16. Gniazdowski, Z.: New interpretation of principal components analysis. Zeszyty Naukowe WWSI 16(11), 43–65 (2017). https://doi.org/10.26348/znwwsi.16.43
17. Everitt, B.S., Landau, S., Leese, M., Stahl, D.: Cluster Analysis. Wiley, London (2011)
18. Rousseeuw, P.J.: Silhouettes: A graphical aid to the interpretation and validation of cluster analysis. J. Comput. Appl. Math. 20(1), 53–65 (1987). https://doi.org/10.1016/0377-0427(87)90125-7
19. Hernández, Á., David, M.: Paper-Complexity-Mexican-geoelectoral-system. GitHub (2021) https://github.com/MDAlvarezH/Paper-Complexity-Mexican-geoelectoral-system
20. Hintze, J.L., Nelson, R.D.. Violin plots: A box plot-density trace synergism. The American Statistician 52(2), 181–184 (1998)
21. Pedregosa, F., Varoquaux, G., Gramfort, A., Michel, V., Thirion, B., Grisel, O., ...Blondel, M.: Scikit-learn: Machine learning in Python. JMLR 12, 2825–2830 (2011). https://scikit-learn.org/stable/auto_examples/index.html

A Spatial Analysis of Drug Dealing in Mexico City

Shaní Alvarez Hernández

1 Introduction

Drug trafficking in Mexico has been the subject of several studies that aim to understand the dynamics taking place within the supply chain of drug markets, including the production, distribution, commercialization, and consumption of drugs [1]. Several supply chain frameworks have been proposed to analyze it [1, 2], but because of its scale, understanding drug trafficking at the global level remains a complex task due to the variability found in illegal markets, which depends on region, drug type, and commercial activity [3].

The importance of understanding this illegal market has been expressed in several government and non-governmental organization's reports [2, 4, 5], academic studies, and journalistic works to demonstrate ways in which the authorities, law enforcement, civil society organizations, and academia can design effective policies, combat and prevention strategies, as well as programs to reduce the supply and demand of drugs.

Findings [5] indicate that despite not being among the countries with the highest levels of reported drug consumption, Mexico has seen a concerning increase in recent years, characterized by early drug use among youths and a lack of addiction treatment programs, especially for populations such as youths in prison.

A review of the literature revealed that studies about the phenomenon of drug trafficking in Mexico have been, in their majority, scoped at the state or regional level. More specifically, studies about drug trafficking in Mexico City are scarce, in part due to a lack of historical, disaggregated, and updated official data on criminal groups and their modus operandi, drug users, drug types and prices, and drug seizures. As a result, these studies have relied on data obtained through journalistic works, field work, and published statistics on crime reports.

S. Alvarez Hernández (✉)
Justice Program, UNODC Mexico, Mexico City, Mexico
e-mail: shanieneida@gmail.com

I. Antoniano-Villalobos et al. (eds.), *Interdisciplinary Statistics in Mexico*,
Springer Proceedings in Mathematics & Statistics 397,
https://doi.org/10.1007/978-3-031-12778-6_2

Consequently, it is at the retail level, referred to as drug dealing or "narcomenudeo," that there is a gap in the official statistics and in the obtainable conclusions to help develop evidence-based policies and strategies against drug selling. According to the General Health Law of Mexico [6], "narcomenudeo" includes the trade, supply, purchase, acquisition, or possession of narcotics and is divided into (1) possession for the purpose of sale, trade, and supply, and (2) simple possession for personal and immediate use for non-commercial purposes.

Information from crime reports [7] indicate that Mexico City is one of the main federal entities exhibiting an increase in crime reports, as evidenced in data from 2015 to 2019. Particularly, Mexico City has had a surge of drug dealing reports, increasing 127.6% from 2017 to 2018, thus becoming the eighth federal entity with the highest rate per 100,000 inhabitants in 2019 and the ninth in 2020 [7]. This situation exposes the need to identify the geographical areas in which this problem is most prominent, as well as those in which drug dealing geographically associates with other crimes so that local authorities, law enforcement, and other relevant stakeholders are able to make informed decisions.

In this context of Mexico City's market for illegal drugs, the aim of this paper is to study the spatial distribution of drug dealing for the purpose of sale at the neighborhood level from 2019 to 2020, and to identify neighborhoods that constitute clusters of this criminal activity. It also aims to analyze the spatial relationship between drug dealing for sale purposes and the following crimes associated to organized crime: illicit carrying of firearms, intentional homicide, extortion, kidnapping, and drug dealing (simple possession).

A geographic and visual assessment is intended to compare the spatial distributions of these clusters in both years and identify any pattern changes, especially in light of the resilience of the drug trafficking market with respect to the changes brought about by the pandemic in 2020 [8]. The results will then be contrasted with the findings of the discussed literature.

Therefore, the following research questions will be explored:

1. Is reported drug dealing for sale purposes spatially correlated at the neighborhood level? If so, in which boroughs and neighborhoods does the spatial distribution show clustering (hotspots) in 2019 and 2020?
2. Is reported drug dealing for sale purposes spatially correlated at the neighborhood level to other spatially lagged crimes associated to organized crime, such as illicit carrying of firearms, intentional homicide, extortion, kidnapping, and drug dealing (simple possession)? If so, in which boroughs and neighborhoods does the spatial distribution show clustering (hotspots) in 2019 and 2020?
3. Are there any identifiable changes in the spatial distribution of reported drug dealing for sale purposes from 2019 to 2020?

2 Background

2.1 Drug Dealing as Part of Organized Crime

Under the Organized Crime Convention, transnational organized crime is defined in broad terms as any serious offense committed by three or more people with the purpose of obtaining some material gain [4]. In Mexico, according to the Federal Law against Organized Crime [9], this refers to crimes committed by groups of people that conspire together to carry out an illicit business. There have been changes in meaning throughout history, but they mostly fall into one of two notions: as criminal activities conducted by organizations with profit as the main goal, or as systematic criminal activities run by stable organizations [10].

Considering the revenue attained at a local scale, drug dealing falls not only into the first category, but also the second one if we consider it as one of the last links in the supply chain of drug trafficking in Mexico. Drug dealing is thus differentiated from high-scale trafficking in the sense that the buyers are the consumers of the product [11].

While the categorization of crimes and their relation to organized crime varies according to the laws, policies, and treaties followed by each country, in the case of Mexico and according to the definition of the Federal Law against Organized Crime [9], these include crimes against health in the form of drug dealing, collection and arms trafficking, human trafficking, robbery or theft of vehicles, kidnapping, among others.

In addition to the crimes included in the Law, organized criminal groups tend to enforce control over a territory by means of violence, threats, or by offering "protection" services [12]. One such case is extortion, especially to businesses, as it fulfills an illegal taxing function regulated by criminal groups. Incidents involving "derecho de piso" [13, 14] have been documented in Mexico City by media, for example.

Moreover, Vilalta et al. [15] express how there cannot be a disconnection between the study of homicidal violence and drug dealing crimes, specifically in Mexico City. Increases in homicide levels have been associated to destabilizations among criminal groups due to conflicts and changes related to drug trafficking, as was the case during the Mexican War on Drugs [12, 16]. Although Mexico City's homicide rate per 100,000 inhabitants decreased from 2019 (15.2) to 2020 (12.3), in the span of five years (2015–2020) there was an increase of 42.4% in intentional homicide reports [7].

2.2 Research Approaches

In the examined literature, drug dealing has been studied through the lens of multiple criminological theories. Rational choice theory [17], routine activity theory [18], and social disorganization theory [19] are some of the approaches covered by researchers, suggesting that the commission of drug dealing results from the conjunction of high-reward, low-risk choices from both buyer and seller, plus the characteristics of the environment that favor this illegal trade.

As an example, Eck [20] proposed a model to analyze the geographical distribution of illicit markets, suggesting that there are two solutions to the problem of finding a suitable place to sell/buy drugs. The first solution is to make use of the seller's/buyer's social network, while the second is to use routine activities to find these suitable locations.

However, in the case of the neighborhood Desarrollo Urbano Quetzalcóatl, located in Mexico City's Iztapalapa borough, Cisneros [21] documented how any place is potentially viable to sell and buy drugs such as marihuana, cocaine, and crack. More specifically, these places were often grocery stores, houses, apartments, or housing units.

Also in Mexico City, Zamudio [22] consider drug dealing as a network of commercial relations that brings together the seller and buyer. Three modalities were found to describe the selling points at which they meet: fixed (grocery stores or "tienditas," other small business premises), semi-fixed (public spaces and events), and ambulatory (on the street). Drug dealers opt for using fixed selling points in marginalized urban areas.

Despite models such as Eck's, which provide a theoretical framework to study the spatial patterns of illicit markets, the previous examples indicate the need to empirically analyze the phenomenon of drug dealing within the social and the cultural context of Latin American countries. More specifically, the case of drug dealing in Mexico City, as it is a study area that remains underdeveloped.

The spatial study of crime makes it possible to identify the characteristics and interrelations of the places and behaviors of the actors involved, such as drug buyers and sellers. The geographical variability of criminal behavior can be subsequently quantified and predicted [23].

In this context, spatial data analyses have been published to analyze the distribution of crime incidents and their relationship with other environmental factors found in the study area. For example, by using crime data from the city of Chicago, USA, Barnum et al. [24] applied Risk Terrain Modeling to analyze possible environmental risk factors associated to drug dealing incidents.

In an undisclosed city in China, Chen et al. [25] also studied possible environmental and sociodemographic risk factors in drug dealing by using the Poisson regression for spatial heterogeneity. More broadly, Blanco [26] used kernel density and spatial autocorrelation to analyze crime density in the city of Formosa, Argentina. Vilalta [23] analyzed the spatial patterns and time series of vehicle theft in Mexico City.

More recently, Vilalta et al. [15] studied the spatial diffusion of homicide in vulnerable populations at the neighborhood level in Mexico City.

Additionally, in the Cuauhtémoc borough of Mexico City, Vilalta [27] showed that drug dealing tends to geographically cluster (hotspots) in places where commercial activity, local socioeconomic issues, and political negligence are present. Evidence suggests that drug dealing is geographically static or strongly congregated in certain neighborhoods, that is, offenders rarely relocate to other places to sell drugs.

3 Methodology

Mexico City is composed of 16 boroughs (see Fig. 1 for their exact locations) containing a varied number of neighborhoods, which were selected as areal units of analysis. A total of $N = 1{,}808$ neighborhoods were considered, corresponding to the georeferenced open dataset provided by the Mexico City Government.

As a result of this selection, the variables of interest were aggregated at the neighborhood level for the purposes of this analysis. This choice is justified, however, since larger units such as boroughs would conceal the local area effects caused by the spatial variability of crime patterns [28, 29], while a choice of smaller units such as

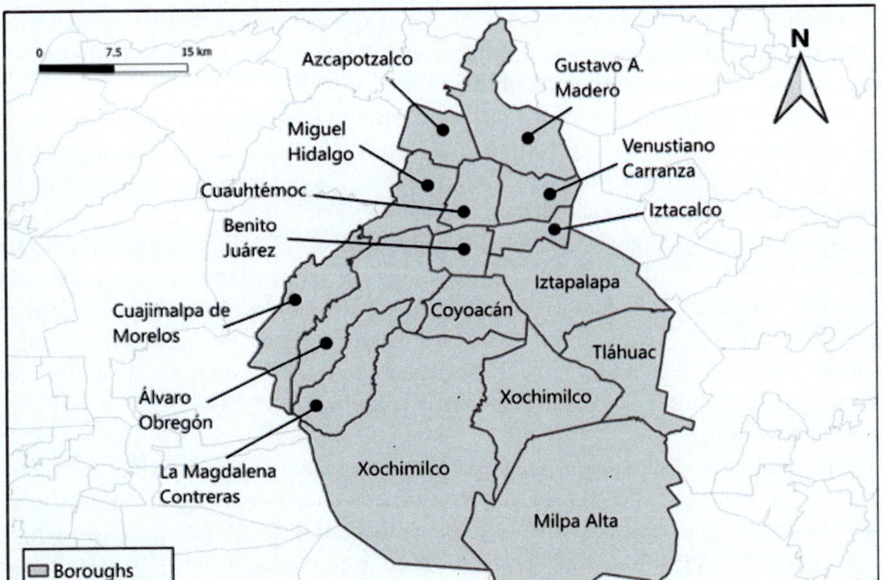

Fig. 1 Mexico City's 16 boroughs

census tracts or standardized grids would leave aside the social context and influence of neighborhoods in social processes such as crime [30]. Also, given the documented endemic and static qualities of drug dealing in Mexico City's neighborhoods [27], these represent suitable units of analysis.

The variable of interest is the number of reports on drug dealing for sale purposes per neighborhood Y_n, where n represents a neighborhood. The crime reports were taken from the open dataset provided by the General Attorney's Office of Mexico City [31]. As a first exploratory approach, only data from the years 2019 and 2020 were selected to analyze the differences in the spatial distribution.

It is worth noting that while crime reports are one of the most common means by which crime is quantified, there are limitations to their use. Crime reports constitute observational discrete data that is subject to errors (e.g., wrong georeferentiation or crime typification). They also constitute a small segment of the total crimes that are actually committed, since most crimes are not reported to the police, and therefore, are not visible in any crime statistics [32].

GIS software such as QGIS version 3.10.1-A Coruña and GeoDa version 1.14.0 were used to process the georeferenced data, apply methods of exploratory data analysis, measure spatial correlation, and subsequently produce mapping representations for visual analysis.

To suitably represent the spatial distribution of drug dealing for sale purposes observations in discrete spatial units, in this case neighborhoods, the data were classified using Jenks natural breaks. With this optimization algorithm, data can be grouped in classes by setting break points that maximize the variance between classes while reducing internal class variance [33]. Choropleth maps allow to visualize the distribution resulting from the classification.

The inherent aggrupation of the data (the patterns) is of interest; therefore, compared to other classification methods (e.g., equal intervals, quantile, k-means, box plot, and standard deviation), Jenks natural breaks was chosen because it emphasizes the natural groupings of the data and clusters extreme values [34]. In this case, five classes were computed; however, tests can be run with less or more breaks to determine the appropriate number of classes for any mapped visual representation.

Moreover, to compare the spatial distribution between years and further highlight possible outliers, a box map was used to visually represent the equivalent of a histogram, except with an added spatial component. In this case, outliers are a function of the hinge or cutoff value [35], which is a multiple, in this case $h = 3$, of the interquartile range.

Global Moran's I statistic provides a measure of spatial autocorrelation, and inference on this statistic is based on a null hypothesis of spatial randomness [36]. However, this statistic is representative of the overall spatial distribution, and as such, it does not provide information about the location of possible clusters [37]. A positive value implies that there exists overall clustering of similar values, a negative value indicates an overall clustering of dissimilar values, and lastly, a value of zero means that there is no spatial autocorrelation [38].

Anselin [39] proposed a revision of the global measure that could analyze the local variations in spatial autocorrelation, where local indicators of spatial association (LISA) measure the extent of significant spatial clustering around an observation. The statistic satisfies two requirements: the LISA for an observation specifies the extent of significant spatial clustering of similar values around that observation, and the sum of all LISA per observation is proportional to its global counterpart.

A LISA for a variable y_i observed at location i is thus defined as [39]

$$L_i = f(y_i, y_{J_i}) \tag{1}$$

where f is a function and y_{J_i} are the observed values or deviations from the mean in the neighborhood J_i of location i.

In this respect, Local Moran's I is a LISA statistic that provides a way to identify local clusters and spatial outliers [38]. For an observation at location i, Local Moran's I statistic can be defined as [39]

$$I_i = z_i \sum_j w_{ij} z_j \tag{2}$$

in which the observations $z_j = y_i - \overline{y}$ (where $\overline{y}$ is the mean of variable y) are deviations from the mean, w_{ij} are the weights, and the summation over j for n observations includes only the neighboring locations.

Following the previous equation, Moran's I statistic can then be defined as [39]

$$I = \sum_i \frac{I_i}{S_0(\sum_i z_i^2/n)} \tag{3}$$

where $S_0 = \sum_i \sum_j w_{ij}$.

Once Local Moran's I indexes are calculated per location, significance needs to be assessed in order to identify local clusters. In this case, the obtention of pseudo significance levels is based on a conditional permutation method, where the variable y_i for location i is fixed while the rest of the values are randomly permuted over all the other locations, resulting in an empirical distribution function [39]. This approach is selected due to the inherent problems in evaluating the distribution of Local Moran's I statistic given that it is not asymptotically normally distributed [40].

Therefore, since Local Moran's I is a relative measure, it is only when the value is interpreted within the context of the pseudo p-value at each location that such location can represent a statistically significant cluster [37]. Therefore, the neighborhoods showing a high-high clustering and a significance level of 99% or less are considered statistically significant and representative of hotspots.

Local Moran's I positive value for a location indicates either a high-high cluster (hotspot) where locations with high values gather around others with high values or a low-low cluster (coldspot) where locations with low values gather around others with low values. On the other hand, a negative value indicates clustering of neighborhoods with dissimilar values (high-low or low-high) [37].

Bivariate Moran's I statistic extends the idea of spatial correlation to the association between the variable of interest and the spatial lag of the second variable being analyzed, as opposed to measuring the association between the variable of interest and its own spatial lag, as seen in the univariate case [41]. Similarly to its univariate counterpart, for an observation at location i, bivariate Local Moran's I statistic [42] is defined as

$$I_i^B = cy_i \sum_j w_{ij} v_j \tag{4}$$

where w_{ij} are the weights, v_j the spatially lagged variable, $\sum_j w_{ij} y_j$ the average of the neighboring values for this variable (its spatial lag), and c a constant scaling factor. Lastly, the summation over j for n observations includes only the neighboring locations.

Despite the common use of this statistic to measure the spatial association between two different variables, the actual spatial correlation between them can be overestimated since it ignores in-place correlation [38, 42]. For this particular case, the statistic measures the relationship between drug dealing for sale purposes and each of the following spatially lagged variables: illicit carrying of firearms, intentional homicide, extortion, kidnapping, and drug dealing (simple possession).

To sum up, univariate Local Moran's I statistic provides a way of identifying significant spatial variation of crime reports between a neighborhood and its adjacent ones, allowing to assess the spatial clustering of neighbors with similar characteristics, which would not be possible if global statistics were measured only. For the calculation of Global and Local Moran's I statistics, contiguity-based spatial weights were calculated using the queen contiguity criterion of the first order, which defines neighbors that share edges and vertices [38].

4 Results

There were a total of $Y_N^{2019} = 1{,}032$ observations (georeferenced reports on drug dealing for sale purposes) in 2019, and $Y_N^{2020} = 1{,}179$ in 2020. See Table 1 in the Appendix, which shows the number of reports per crime for 2019 and 2020.

As shown in Fig. 2, the choropleth map displays the distribution of reported drug dealing for sale purposes during 2019. The darker shades on the map represent the neighborhoods classified as the ones with extreme values. These neighborhoods are located in the Cuauhtémoc, Venustiano Carranza, Coyoacán, Alvaro Obregón, and Iztapalapa boroughs.

In 2020, the distribution of reported drug dealing for sale purposes displays neighborhoods with a high number of drug dealing for sale purposes reports in the Cuauhtémoc, Venustiano Carranza, Miguel Hidalgo, Gustavo A. Madero, and Iztapalapa boroughs, as seen in Fig. 3.

A comparison of the spatial distribution between 2019 and 2020, as shown in the map boxes in Fig. 4, shows that the outlier neighborhoods underwent a shift in location, as the neighborhoods in the center-south are no longer outliers by 2020, while

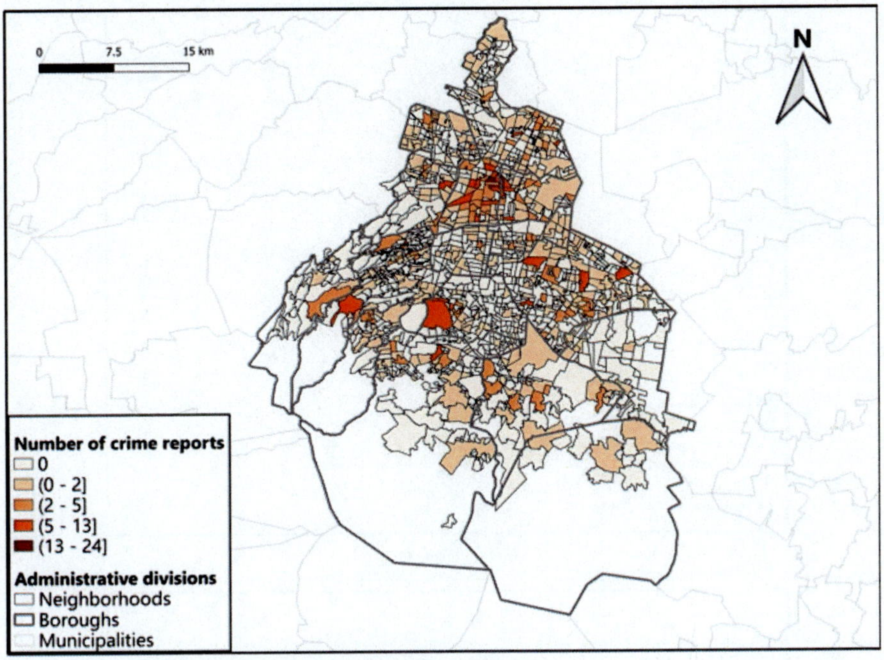

Fig. 2 Distribution of reported drug dealing for sale purposes in Mexico City, 2019

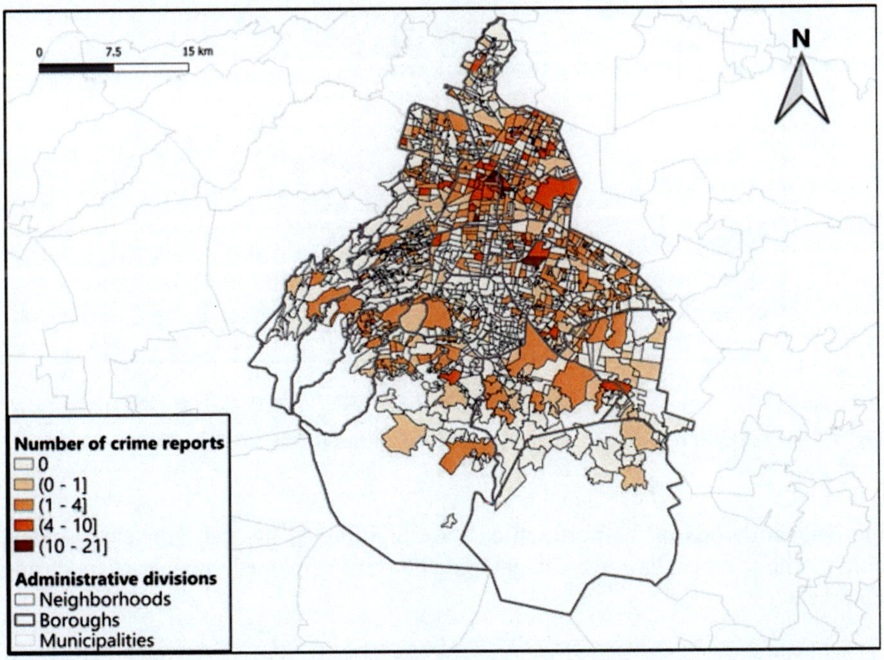

Fig. 3 Distribution of reported drug dealing for sale purposes in Mexico City, 2020

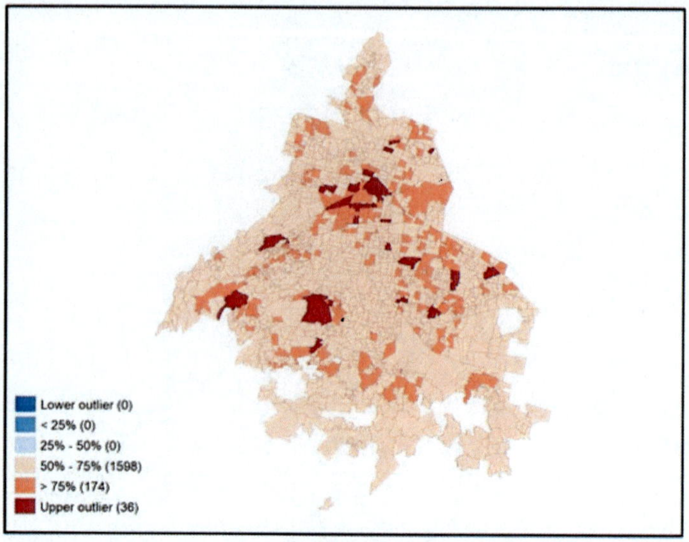

(a) 2019

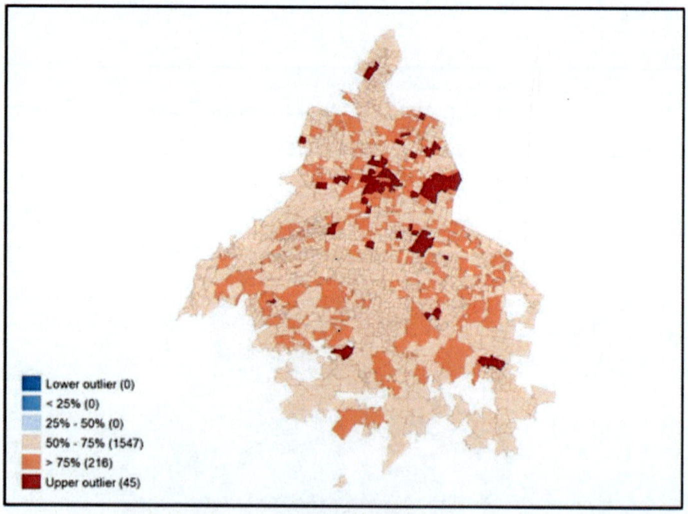

(b) 2020

Fig. 4 Map boxes of reported drug dealing for sale purposes in Mexico City, 2019 and 2020

the neighborhoods in the northern east are highlighted instead. However, several neighborhoods located in the Cuauhtémoc borough remained upper outliers during both years.

4.1 Univariate Global and Local Moran's I

By measuring global spatial autocorrelation, Moran's global index for 2019 is $I = 0.293$, and for 2020 it is $I = 0.28$, which indicates a weak clustering pattern of neighborhoods with similar values. Local clusters can be identified by calculating the coefficients of local Moran's I per neighborhood. In 2019, neighborhoods denoted as high-high are located mainly in the Cuauhtémoc borough, as well as in Venustiano Carranza and Coyoacán. By 2020, both Cuauhtémoc and Venustiano Carranza hold most of the hotspots, followed by the Miguel Hidalgo and Iztapalapa boroughs (see Fig. 5, Tables 2 and 8 in the Appendix for a list of boroughs that contain the identified hotspots, as well as some of these neighborhoods).

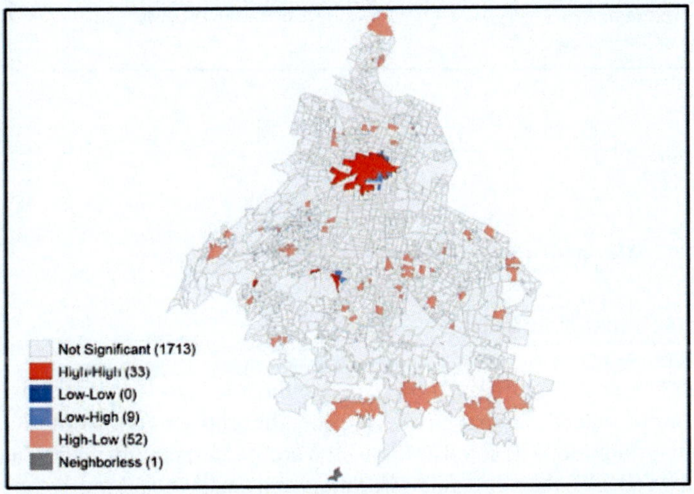

(a) 2019

Fig. 5 Univariate local Moran's I cluster map: drug dealing for sale purposes ($p = 0.01$ or less), 2019–2020

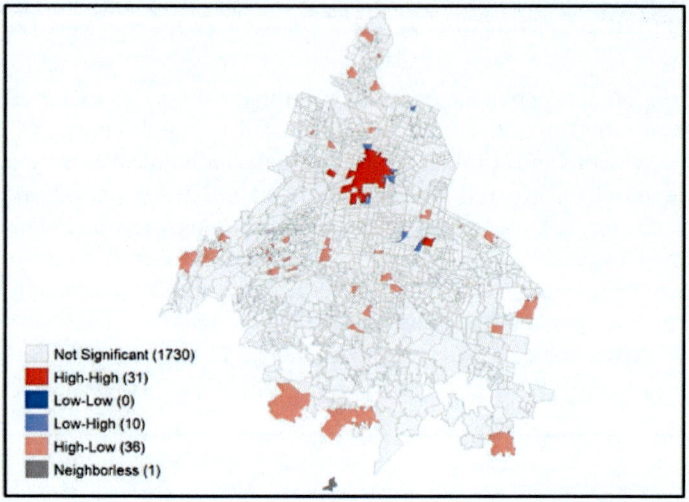

(b) 2020

Fig. 5 (continued)

4.2 Bivariate Global and Local Moran's I

4.2.1 Intentional Homicide

Moran's global index for 2019 is $I^B = 0.169$, and $I^B = 0.164$ for 2020, which denote an almost non-existent clustering pattern of neighborhoods with similar values. For 2019, the neighborhoods denoted as high-high are located mainly in the Cuauhtémoc borough, as well as in the Venustiano Carranza and Gustavo A. Madero boroughs. In 2020, both Cuauhtémoc and Venustiano Carranza hold the most hotspots, followed by Iztapalapa (see Fig. 6, Tables 3 and 9 in the Appendix for a list of boroughs that contain the identified hotspots, as well as some of these neighborhoods).

4.2.2 Illicit Carrying of Firearms

Moran's global index for 2019 is $I^B = 0.139$, and $I^B = 0.110$ for 2020, which denote an almost non-existent clustering pattern of neighborhoods with similar values. For 2019, the neighborhoods denoted as high-high are located mainly in the Cuauhtémoc borough, as well as in the Miguel Hidalgo borough. As for 2020, both Cuauhtémoc and Iztapalapa hold the most hotspots (see Fig. 7, Tables 4 and 10 in the Appendix for a list of boroughs that contain the identified hotspots, as well as some of these neighborhoods).

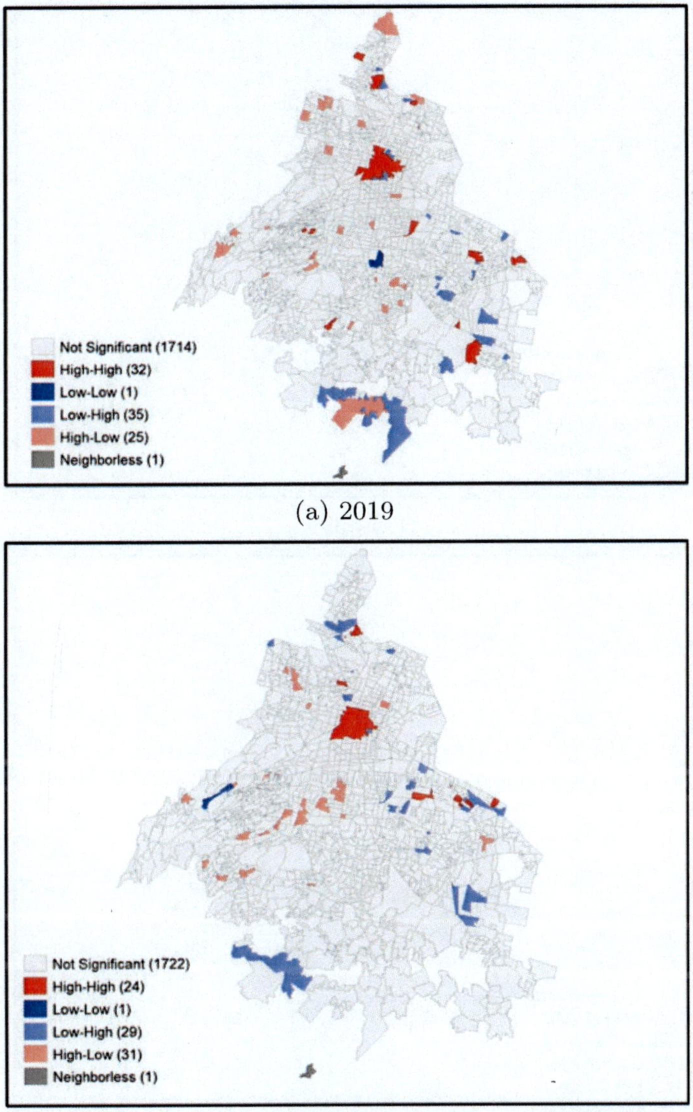

(a) 2019

(b) 2020

Fig. 6 Bivariate local Moran's I cluster map: drug dealing for sale purposes and intentional homicide ($p = 0.01$ or less), 2019–2020

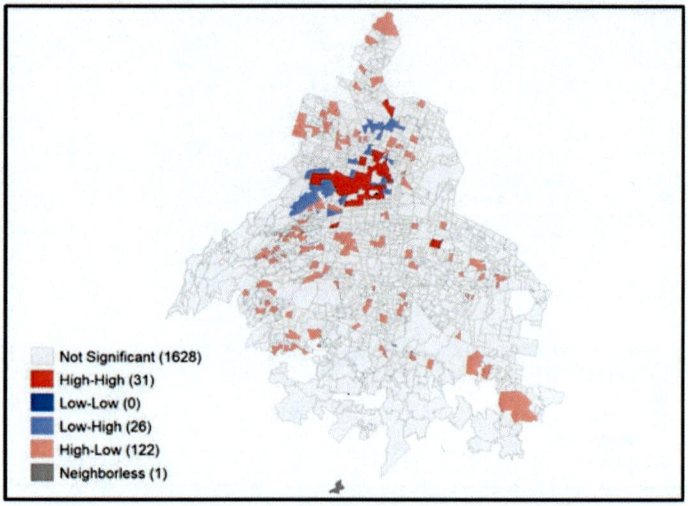

(a) 2019

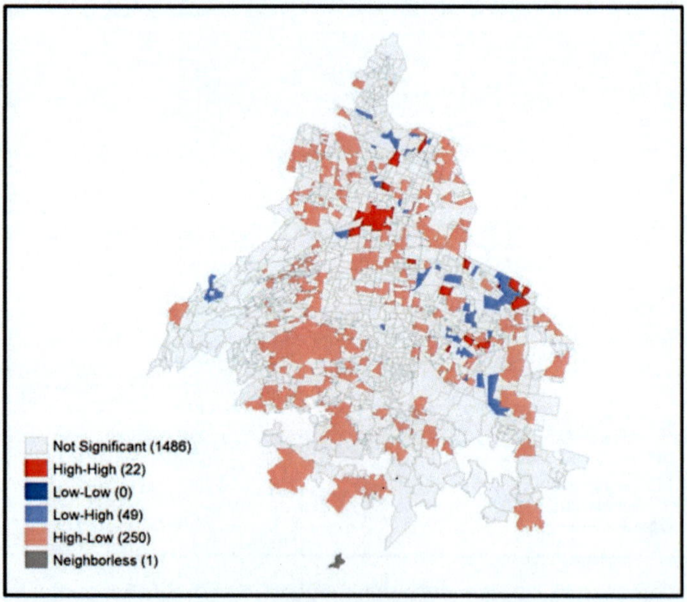

(b) 2020

Fig. 7 Bivariate local Moran's I cluster map: drug dealing for sale purposes and illicit carrying of firearms ($p = 0.01$ or less), 2019–2020

4.2.3 Extortion

Moran's global index for 2019 is $I^B = 0.139$ and $I^B = 0.102$ for 2020, once again denoting an almost non-existent clustering pattern of neighborhoods with similar values. For 2019, the neighborhoods denoted as high-high are located primarily in the Cuauhtémoc borough. As for 2020, this same borough holds the most hotspots (see Fig. 8, Tables 5 and 11 in the Appendix for a list of boroughs that contain the identified hotspots, as well as some of these neighborhoods).

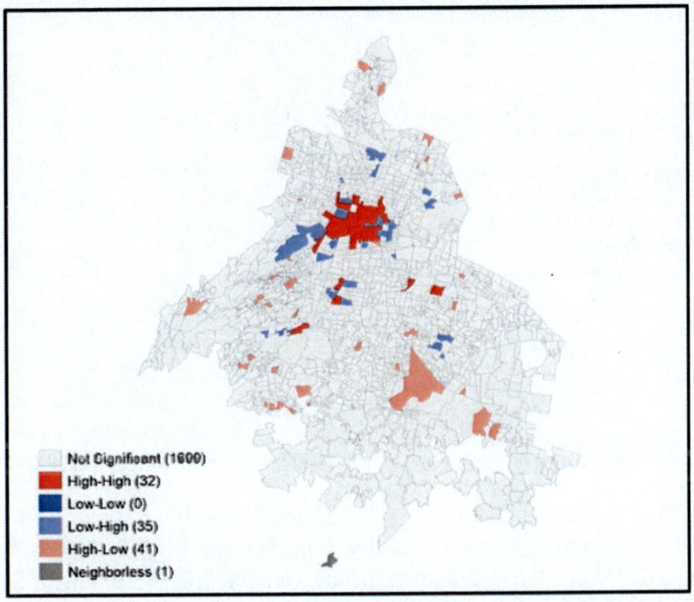

(a) 2019

Fig. 8 Bivariate local Moran's I cluster map: drug dealing for sale purposes and extortion ($p = 0.01$ or less), 2019–2020

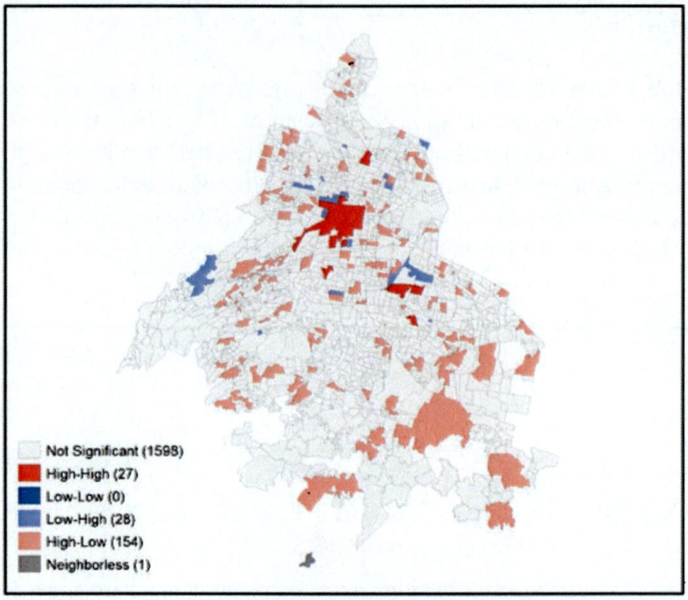

(b) 2020

Fig. 8 (continued)

4.2.4 Kidnapping

Moran's global index for 2019 is $I^B = 0.062$ and $I^B = 0.025$ for 2020, which indicate a random clustering pattern. Locally, however, the neighborhoods denoted as high-high in 2019 are located mainly in the Cuauhtémoc borough. In 2020, both Iztapalapa and Benito Juárez contain all the hotspots (see Fig. 9 and Table 6 in the Appendix for a list of boroughs that contain the identified hotspots).

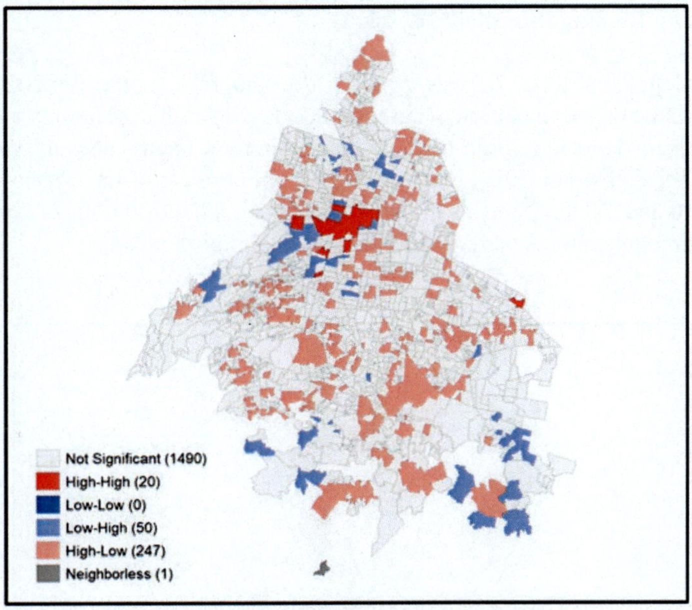

(a) 2019

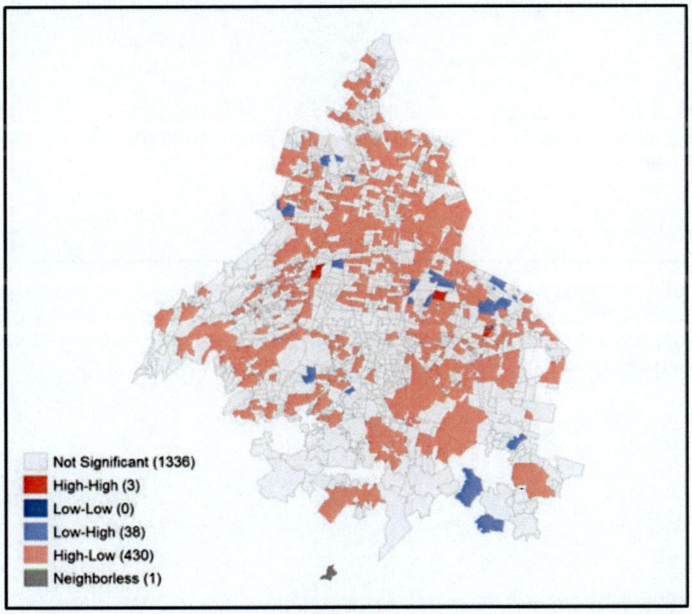

(b) 2020

Fig. 9 Bivariate local Moran's I cluster map: drug dealing for sale purposes and kidnapping ($p = 0.01$ or less), 2019–2020

4.2.5 Drug Dealing (Simple Possession)

Moran's global index for 2019 is $I^B = 0.273$, and $I^B = 0.280$ for 2020, which denote a weak clustering pattern of neighborhoods with similar values. For 2019, the neighborhoods denoted as high-high are located in the Cuauhtémoc and Venustiano Carranza boroughs. For 2020, both boroughs contain most of the hotspots as well (see Fig. 10 and Tables 7 and 12 in the Appendix for a list of boroughs that contain the identified hotspots, as well as some of these neighborhoods).

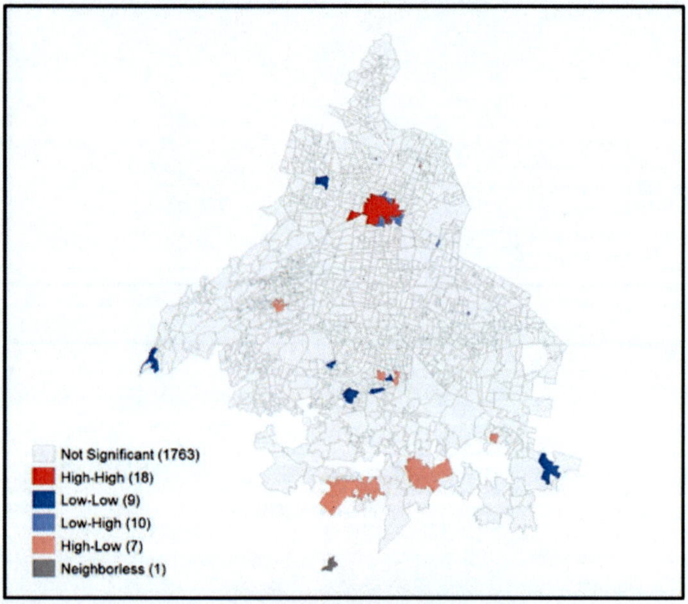

(a) 2019

Fig. 10 Bivariate local Moran's I cluster map: drug dealing for sale purposes and drug dealing (simple possession) ($p = 0.01$ or less), 2019–2020

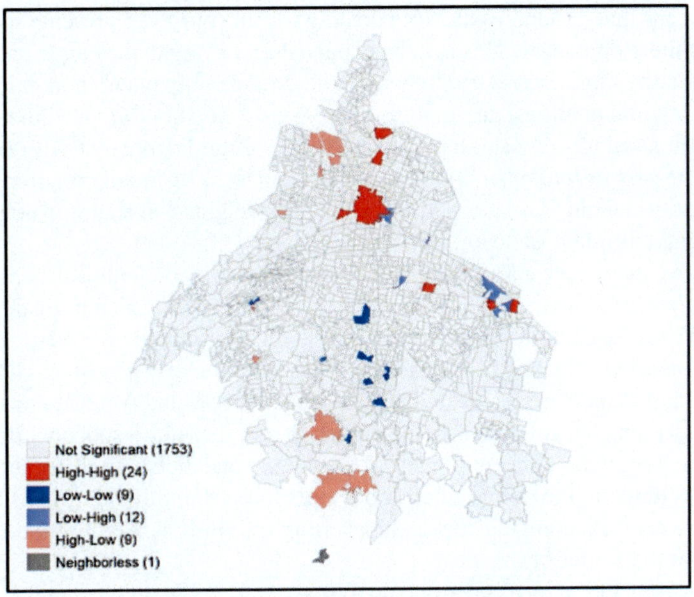

(b) 2020

Fig. 10 (continued)

5 Discussion and Conclusions

This study presented an analysis of the spatial patterns and hotspots of drug dealing for sale purposes in Mexico City during 2019 and 2020 by using measures of spatial association, such as Global and Local Moran's I statistics. It also examined the existence of a correlation between drug dealing for sale purposes and the spatially lagged crimes considered part of the activities undertaken by organized crime in an effort to identify the neighborhoods and boroughs where hotspots are present.

The results obtained from the Global Moran's I univariate index with respect to drug dealing for sale purposes showed that there was a weak clustering pattern of neighborhoods with similar values in 2019 ($I = 0.293$) and 2020 ($I = 0.28$). The indexes did not present considerable changes from one year to another. Similarly, the results obtained from the bivariate global measures indicated a weak clustering pattern in the case of drug dealing for sale purposes and the spatially lagged variable of drug dealing (simple possession), and likewise, the indexes did not present considerable changes from one year to the next. In all other instances, there was essentially no spatial correlation, or it was extremely weak, and once again the difference in index values between years did not denote sizable changes.

This means that, though weak, drug dealing for sale purposes presents spatial correlation at the global level. This is to be expected according to the empirical findings discussed in the literature review by which the drug dealing phenomenon in Mexico City has endemic qualities, and therefore, there exist neighborhoods with neighbors that contain similarly reported crime events. The same is apparent with respect to the bivariate case described previously, as there is a weak spatial correlation between reported drug dealing for sale purposes and the weighted average of neighboring drug dealing (simple possession) reports.

As for the local level analysis of neighborhoods, the coefficients of Local Moran's I allow to interpret the results according to their local statistical significance. In the univariate case, high-high clusters with a 99% level of confidence were identified. With respect to drug dealing for sale purposes, in 2019 the majority of neighborhoods identified as hotspots were located in the Cuauhtémoc and Venustiano Carranza boroughs, which concentrated 81.8% of the total number of hotspots, while in 2020 these same boroughs concentrated 93.5% of the total hotspots. This means that there was a reduction in the number of hotspots located in other boroughs by 2020, indicating a centralization of drug dealing selling activities in neighborhoods located primarily at the center of the city.

With respect to the local bivariate analyses, the identification of high-high clusters with a 99% level of confidence is indicative of the existence of neighborhoods containing reported crimes of drug dealing for sale purposes similar to the weighted average of the neighboring reported crimes for either of the other six crimes. For instance, in the bivariate case of intentional homicide, Cuauhtémoc and Venustiano Carranza stand out as the boroughs that concentrate most of the neighborhoods identified as hotspots for 2019 and 2020. As for drug dealing (simple possession), the same boroughs contain the majority of hotspots in both years.

In the case of illicit carrying of firearms, the Cuauhtemoc and Miguel Hidalgo boroughs concentrate most of the neighborhoods categorized as hotspots in 2019; however, by 2020, Iztapalapa stands out as the second borough containing the majority of hotspots. In the case of extortion, Cuauhtémoc contains most of the hotspots in both years. As for kidnapping, there is an important change in the locations of hotspots, since by 2020 the Iztapalapa and Benito Juárez boroughs substitute Cuauhtémoc as the boroughs with the most hotspots identified.

In line with the previous results, the cluster maps show a reduction of the number of hotspots in all cases, except drug dealing (simple possession), where there is an increase in the number of hotspots in 2020 in comparison with the previous year, showcasing more dispersion over the city. Overall, the results correspond with the examined literature in that drug dealing has endemic and static properties, while also providing new insights into this crime's relationship with the other criminal activities related to organized crime.

To conclude, this research has room for expansion and improvement, not only in the analysis of drug dealing and organized crime, but in the analysis of other crimes that prevail in the public sphere of Mexico City. Other classification methods can be tested with the aim to improve the visualization of crimes in choropleth maps, and other measures of spatial association, such as Geary's C, can be tested and compared with the results obtained with Global and Local Moran's I measures. Lastly, research into drug dealing can go beyond the analysis of spatial autocorrelation and consider regression methods to introduce in the statistical modeling more geographic context, as well as other factors that contribute to the manifestation of these crimes.

Notes and Comments. It is suggested to the reader to see [43] in order to review all Local Moran's I cluster and significance maps, as well as the graphs that provide a visual representation of Global Moran's I index.

Appendix

Table 1 Number of crimes reported in Mexico City, 2019–2020

Type	2019	2020
Drug dealing for sale purposes	1,032	1,179
Drug dealing (simple possession)	5,017	4,187
Intentional homicide	1,344	1,129
Extortion	793	326
Illegal carrying of firearms	420	213
Kidnapping	176	58

Table 2 Number of hotspots per borough, univariate local Moran I of drug dealing for sale purposes, 2019–2020

Borough	Number of hotspots		Variation
	2019	2020	
Alvaro Obregon	2	0	–
Azcapotzalco	0	0	=
Benito Juarez	0	0	=
Coyoacan	3	0	–
Cuajimalpa	0	0	=
Cuauhtemoc	22	23	+
Gustavo A. Madero	0	0	=
Iztacalco	0	0	=
Iztapalapa	0	1	+
La Magdalena Contreras	0	0	=
Miguel Hidalgo	1	1	=
Milpa Alta	0	0	=
Tlahuac	0	0	=
Tlalpan	0	0	=
Venustiano Carranza	5	6	+
Xochimilco	0	0	=

Table 3 Number of hotspots per borough, bivariate local Moran I of drug dealing for sale purposes and intentional homicide, 2019–2020

Borough	Number of hotspots		Variation
	2019	2020	
Alvaro Obregón	2	0	–
Azcapotzalco	0	0	=
Benito Juárez	0	0	=
Coyoacán	0	0	=
Cuajimalpa	0	0	=
Cuauhtémoc	13	12	–
Gustavo A. Madero	4	2	–
Iztacalco	1	0	–
Iztapalapa	3	5	+
La Magdalena Contreras	0	0	=
Miguel Hidalgo	0	0	=
Milpa Alta	0	0	=
Tlahuac	1	0	–
Tlalpan	1	0	–
Venustiano Carranza	6	5	–
Xochimilco	1	0	–

Table 4 Number of hotspots per borough, bivariate local Moran I of drug dealing for sale purposes and illicit carrying of firearms, 2019–2020

Borough	Number of hotspots		Variation
	2019	2020	
Alvaro Obregón	0	0	=
Azcapotzalco	0	0	=
Benito Juárez	1	0	–
Coyoacán	0	0	=
Cuajimalpa	0	0	=
Cuauhtémoc	22	10	–
Gustavo A. Madero	1	2	+
Iztacalco	0	1	+
Iztapalapa	1	9	+
La Magdalena Contreras	0	0	=
Miguel Hidalgo	5	0	–
Milpa Alta	0	0	=
Tlahuac	0	0	=
Tlalpan	0	0	=
Venustiano Carranza	1	0	–
Xochimilco	0	0	=

Table 5 Number of hotspots per borough, bivariate local Moran I of drug dealing for sale purposes and extortion, 2019–2020

Borough	Number of hotspots		Variation
	2019	2020	
Alvaro Obregón	1	0	–
Azcapotzalco	0	0	=
Benito Juárez	2	1	–
Coyoacán	0	0	=
Cuajimalpa	0	0	=
Cuauhtémoc	23	21	–
Gustavo A. Madero	0	1	+
Iztacalco	0	0	=
Iztapalapa	2	3	+
La Magdalena Contreras	0	0	=
Miguel Hidalgo	3	1	–
Milpa Alta	0	0	=
Tlahuac	0	0	=
Tlalpan	0	0	=
Venustiano Carranza	1	0	–
Xochimilco	0	0	=

Table 6 Number of hotspots per borough, bivariate local Moran I of drug dealing for sale purposes and kidnapping, 2019–2020

Borough	Number of hotspots		Variation
	2019	2020	
Alvaro Obregón	0	0	=
Azcapotzalco	0	0	=
Benito Juárez	1	1	=
Coyoacán	0	0	=
Cuajimalpa	0	0	=
Cuauhtémoc	15	0	–
Gustavo A. Madero	0	0	=
Iztacalco	0	0	=
Iztapalapa	1	2	+
La Magdalena Contreras	0	0	=
Miguel Hidalgo	3	0	–
Milpa Alta	0	0	=
Tlahuac	0	0	=
Tlalpan	0	0	=
Venustiano Carranza	0	0	=
Xochimilco	0	0	=

Table 7 Number of hotspots per borough, bivariate local Moran I of drug dealing for sale purposes and drug dealing (simple possession), 2019–2020

Borough	Number of hotspots		Variation
	2019	2020	
Alvaro Obregón	0	2	+
Azcapotzalco	0	0	=
Benito Juárez	0	0	=
Coyoacán	0	0	=
Cuajimalpa	0	0	=
Cuauhtémoc	11	14	+
Gustavo A. Madero	0	0	=
Iztacalco	0	0	=
Iztapalapa	0	3	+
La Magdalena Contreras	0	0	=
Miguel Hidalgo	0	0	=
Milpa Alta	0	0	=
Tlahuac	0	0	=
Tlalpan	0	0	=
Venustiano Carranza	7	5	–
Xochimilco	0	0	=

Table 8 Examples of hotspots identified in both 2019 and 2020, univariate local Moran's I, drug dealing for sale purposes

Borough	Neighborhoods
Cuauhtémoc	Tabacalera
	Centro I
	Morelos I
	Doctores I
Venustiano Carranza	Centro II
	Amp. Penitenciaria
	Morelos II

Table 9 Examples of hotspots identified in both 2019 and 2020, bivariate local Moran's I, drug dealing for sale purposes and intentional homicide

Borough	Neighborhoods
Cuauhtémoc	Morelos I
	Morelos III
	Centro VIII
	Guerrero II
Venustiano Carranza	Centro I
	Morelos II
	Amp. Penitenciaria

Table 10 Examples of hotspots identified in both 2019 and 2020, bivariate local Moran's I, drug dealing for sale purposes and illicit carrying of firearms

Borough	Neighborhoods
Cuauhtémoc	Guerrero IV
	Centro VIII
	Roma Norte II
	Tabacalera
Iztapalapa	Guadalupe del Moral

Table 11 Examples of hotspots identified in both 2019 and 2020, bivariate local Moran's I, drug dealing for sale purposes and extortion

Borough	Neighborhoods
Cuauhtémoc	Juárez
	Centro VIII
	Roma Norte II
	Doctores I

Table 12 Examples of hotspots identified in both 2019 and 2020, bivariate local Moran's I, drug dealing for sale purposes and drug dealing (simple possession)

Borough	Neighborhoods
Cuauhtémoc	Tabacalera
	Guerrero IV
	Centro IV
	Morelos III
Venustiano Carranza	Morelos II
	Amp. Penitenciaria

References

1. RAND Europe.: Understanding illicit drug markets, supply-reduction efforts, and drug-related crime in the European Union. RAND Corporation, Cambridge (2010)
2. UNODC.: World Drug Report 2021. United Nations, Vienna (2021)
3. Reuter, P.: Drug Markets and Organized Crime. The Oxford Handbook of Organized Crime, 359–380. Oxford University Press, New York (2014)
4. UNODC.: The Globalization of Crime. A Transnational Organized Crime Threat Assessment. United Nations, Vienna (2010)
5. Comisión Nacional contra las Adicciones.: Informe sobre la Situación del Consumo de Drogas en México y su Atención Integral 2019. Gobierno de México, Mexico City (2019)
6. Cámara de Diputados del H. Congreso de la Unión.: Ley General de Salud. (2021). http://www.diputados.gob.mx/LeyesBiblio/pdf_mov/Ley_General_de_Salud.pdf
7. SESNSP.: Incidencia delictiva. Gobierno de México, (2021). https://www.gob.mx/sesnsp/acciones-y-programas/incidencia-delictiva-87005?idiom=es
8. UNODC.: World Drug Report 2021. COVID-19 and Drugs: Impact Outlook. United Nations, Vienna (2021)
9. Cámara de Diputados del H. Congreso de la Unión.: Ley Federal contra la Delincuencia Organizada. (2021). http://www.diputados.gob.mx/LeyesBiblio/pdf/101_200521.pdf
10. Paoli, L., Beken, T.V.: Organized Crime: A Contested Concept. The Oxford Handbook of Organized Crime, 13–31. Oxford University Press (2014)
11. Zamudio, A., Carlos, A.: "Qué es el narcomenudeo?" Un acercamiento etnológico. Revista Liberaddictus (103), 3–7 (2008). http://dx.doi.org/10.30972/geo.17334426
12. UNODC.: Global Study on Homicide 2019. United Nations, Vienna (2019)
13. Infobae.: "Mapa del "cobro de piso" en CDMX: dónde se han reportado extorsiones del crimen organizado." Infobae, (01/12/2020). https://www.infobae.com/america/mexico/2020/12/01/mapa-del-cobro-de-piso-en-cdmx-donde-se-han-reportado-extorsiones-del-crimen-organizado/
14. La Redacción.: "Extorsiones y cobro por derecho de piso aumentan en tres alcaldías de CDMX: Canaco." Proceso, (10/09/2021). https://www.proceso.com.mx/nacional/2021/9/10/extorsiones-cobro-por-derecho-de-piso-aumentan-en-tres-alcaldias-de-cdmx-canaco-271753.html
15. Vilalta, C., Lopez-Ramirez, P., Fondevila, G.: The spatial diffusion of homicide in Mexico City: a test of theories in context. Global Crime 22(3), 222–239 (2021). https://doi.org/10.1080/17440572.2021.1909480
16. Rosen, J.D., Zepeda Martinez, R.: La guerra contra el narcotráfico en México: una guerra perdida. Reflexiones 94(1), 153–168 (2015). https://doi.org/10.15517/rr.v94i1.20889
17. Clarke, R.V., Cornish, D.B.: Modeling Offenders' Decisions: A Framework for Research and Policy. Crime and Justice: An Annual Review of Research, vol. 6 (1985)
18. Cohen, L.E., Felson, M.: Social Change and Crime Rate Trends: A Routine Activity Approach. Am. Sociol. Rev. 44(4), 588–608 (1979)
19. Shaw, C.R., McKay, H.D.: Juvenile Delinquency and Urban Areas. University of Chicago Press, Chicago (1942)
20. Eck, J.E.: A General Model of the Geography of Illicit Retail Marketplaces. Crime and Place: Crime Prevention Studies, vol. 4, pp. 67–93. Police Executive Research Forum, Washington, D.C (1995)
21. Cisneros, J.L.: La neta somos así. Violencia cotidiana, espacio y juventud: el caso de la colonia Desarrollo Urbano Quetzalcóatl, Ciudad de México. Poliíticas de seguridad y entornos violentos en Colombia y México, 181–198. Universidad de Medelliín, Medellín (2018)
22. Zamudio, A., Carlos, A.: Jóvenes en el narcomenudeo: el caso Ciudad de México. URVIO, Revista Latinoamericana de Seguridad Ciudadana 13, 111–123 (2013). https://doi.org/10.17141/urvio.13.2013.1185
23. Vilalta, P., Carlos, J.: El robo de vehículos en la ciudad de México. Patrones espaciales y series de tiempo. Gestión y Política Pública 20(1), 97–139 (2011)

24. Barnum, J.D., Campbell, W.L., Trocchio, S., Caplan, J.M., Kennedy, L.W.: Examining the Environmental Characteristics of Drug Dealing Locations. Crime Delinq. 63(13), 1731–1756 (2017). https://doi.org/10.1177/2F0011128716649735
25. Chen, J., Liu, L., Liu, H., Long, D., Xu, C., Zhou, H.: The spatial heterogeneity of factors of drug dealing: A case study from ZG, China. ISPRS Int. J. Geo-Inf. 9(4) (2020). https://doi.org/10.3390/ijgi9040205
26. Blanco, H.D.: Análisis Espacial de Patrones Delictivos en la Ciudad de Formosa, Argentina. Actas Científicas Congreso Internacional de Geografía - 78° Semana de la Geografía, 43–58 (2017)
27. Vilalta P., Carlos J.: La geografía local del narcomenudeo: patrones, procesos y recomendaciones de política urbana. Estudios Demográficos y Urbanos 24(1), 49–77 (2009). https://doi.org/10.24201/edu.v24i1.1344
28. Weisburd, D., Bruinsma, G.J., Bernasco, W.: Units of Analysis in Geographic Criminology: Historical Development, Critical Issues, and Open Questions in Putting Crime in its Place, vol. 3–31. Springer (2009)
29. Weisburd, D., Groff, E.R., Yang, S.-M.: The Criminology of Place: Street segments and our understanding of the crime problem, Oxford University Press, 40–41 (2012)
30. Tita, G.E., Greenbaum, R.T.: Crime, Neighborhoods, and Units of Analysis in Putting Crime in its Place, pp. 145–170. Springer (2009)
31. Fiscalía General de Justicia (FGJ) de la Ciudad de México.: Carpetas de investigación FGJ de la Ciudad de México. Portal de Datos Abiertos de la CDMX, (2021). https://datos.cdmx.gob.mx/dataset/carpetas-de-investigacion-fgj-de-la-ciudad-de-mexico
32. UNODC.: Comisión Económica de las Naciones Unidas para Europa. Manual Para Encuestas de Victimización. Naciones Unidas, Ginebra (2010)
33. de Smith, M.J., Goodchild, M.F., Longley, P.A.: Geospatial Analysis. A Comprehensive Guide to Principles, Techniques and Software Tools. de Smith, Goodchild, Longley and Associates (2018–2021)
34. ArcGIS.: Data classification methods—ArcGIS Pro Documentation. Esri, (2021). https://pro.arcgis.com/en/pro-app/latest/help/mapping/layer-properties/data-classification-methods.htmESRI_SECTION1_B47C458CFF6A4EEC933A8C7612DA558B
35. Anselin, L.: Basic Mapping. GeoDa, (2021). https://geodacenter.github.io/workbook/3a_mapping/lab3a.htmlthematic-maps-overview
36. Anselin, L.: Global Spatial Autocorrelation. GeoDa, (2020). https://geodacenter.github.io/workbook/5a_global_auto/lab5a.html
37. Anselin, L.: Local Spatial Autocorrelation (1). Geoda, (2020). https://geodacenter.github.io/workbook/6a_local_auto/lab6a.html
38. Lloyd, C.D.: Local Models for Spatial Analysis. CRC Press (2007)
39. Anselin, L.: Local Indicators of Spatial Association—LISA. Geogr. Anal. 27(2), 93–115 (1995) https://doi.org/10.1111/j.1538-4632.1995.tb00338.x
40. Tiefelsdorf, M.: The saddlepoint approximation of Moran's I's and local Moran's Ii's reference distributions and their numerical evaluation. Geogr. Anal. 34(3), 187–206 (2002). https://doi.org/10.1111/j.1538-4632.2002.tb01084.x
41. Anselin, Luc.: Global Spatial Autocorrelation (2): Bivariate, Differential and EB Rate Moran Scatter Plot. Geoda, (2019) https://geodacenter.github.io/workbook/5b_global_adv/lab5b.html
42. Anselin, Luc.: Local Spatial Autocorrelation (3): Multivariate Local Spatial Autocorrelation. Geoda, (2020). https://geodacenter.github.io/workbook/6c_local_multi/lab6c.htmlbivariate-local-moran
43. Hernández, S.A.: Spatial-analysis/drug-dealing-Mexico-City. GitHub, (2021). https://github.com/shani-alvarez/Spatial-analysis/tree/main/drug-dealing-Mexico-City

Bayesian Hierarchical Multinomial Modeling of the 2021 Mexican Election Outcomes with Censored Samples

Michelle Anzarut, Luis Felipe González, and María Teresa Ortiz

1 Introduction

In Mexico, the elections always take place on a Sunday, and the official results are presented to the population a week later. To prevent unjustified victory claims during that period, the electoral authority organizes a quick count the same day of the election. The quick count has become very important. The rapidity and precision of its results foster trust in the population and serve as a tool against fraud.

The quick count consists of selecting a random sample of the polling stations and estimating the percentage of votes in favor of each candidate. The electoral authority creates a committee of specialists in charge of producing the inferences, which are presented by intervals with a confidence level of at least 95%.

The information system starts at 6 p.m. and, every 5 min, receives new sample information. Thus, the system produces a sequence of accumulative partial samples, analyzed with the estimation methods to track the trend of the results. The quick-count results are made public using one of these partial samples, since a significant percentage of the sample takes too long to arrive. The missingness in the partial samples is not at random; examples of variables that affect the arrival time of a polling station are whether they are in urban or rural sections, the weather, or how many registered voters they have. Even more, we will show the probability that a polling station is not reported at a certain time depends heavily on the response we intend to measure. That is the key challenge of the quick count.

M. Anzarut (✉)
Universidad Nacional Autónoma de México, Mexico City, Mexico
e-mail: michelle@sigma.iimas.unam.mx

L. F. González · M. T. Ortiz
Instituto Tecnológico Autónomo de México, Mexico City, Mexico
e-mail: luis.gonzalez@itam.mx

M. T. Ortiz
e-mail: maria.ortiz@itam.mx

I. Antoniano-Villalobos et al. (eds.), *Interdisciplinary Statistics in Mexico*,
Springer Proceedings in Mathematics & Statistics 397,
https://doi.org/10.1007/978-3-031-12778-6_3

Mexico is divided into 32 states where each state has a governor. The governor is the highest representative of the Executive Power in the state, whose mandate extends for 6 yrs. In the elections of June 6, 2021, the positions for governor for 15 states of the country were renewed, among other positions. Our team was in charge of the quick count for the governor's position in four states: Chihuahua, Colima, Nayarit, and Zacatecas. In this paper, to show the process of model building, we use the data of Nayarit. The model we used for these states is a Bayesian multilevel regression, where the dependent variable has a negative binomial distribution. Hence, we call the *negative binomial multilevel model* (denoted as the NBM model).

The outline of the paper is as follows. In Sect. 2, we describe the background of the NBM model. In Sect. 3, we define the NBM model and its estimation procedure. Also, we expose several steps of the Bayesian workflow we used to validate both the model and its estimation. In Sect. 4, we apply the model to the elections of 2021. Finally, in Sect. 5, we give some concluding remarks and future research directions.

2 Background

In this section, we describe the background of the NBM model; we recount other models that have been used in quick counts in Mexico and discuss where the ideas for the NBM model proposal come from. Also, we explain the context of the bias problem, as it is a determining factor in all steps of the workflow.

2.1 Estimation Methods in the 2021 Quick Count

The most widely used method in the 2021 quick count is the combined ratio estimator (see, for example, [8]). To build confidence intervals with this estimator, nonparametric bootstrap techniques, construction of the sampling distribution via simulation, or approximations to a Normal distribution can be used.

When the full sample has arrived (or something close to it), the combined ratio estimator is a powerful technique that provides intervals with small estimation errors in a matter of seconds. However, in this case, the chosen sample design was stratified, and when there are empty strata, or strata that contain only one polling station, the calculation of this estimator variance is indeterminate. Therefore, it is necessary to apply other resources, for example, strata groupings. This type of grouping can induce bias in the variance estimates. Also, the decision of the strata to collapse must be done prior to the selection of the sample.

Another method used in the 2021 quick count was the contribution of [9], a Bayesian parametric model where the number of people in favor of a candidate divided by poll and stratum has a normal distribution. Let y_i be the votes for the candidate in the i-th polling station, then

$$y_i \big|\ {}_i,\ {}_i \sim \mathsf{N}\left(n_i\ {}_i,\ \frac{{}_i}{n_i}\right),\quad \text{for } i = 1, ..., c,$$

where c is the number of polling stations in the stratum, n_i stands for the size of the nominal list (list of registered voters) in the polling station, and ${}_i$ is the precision.

The authors use the improper prior

$$p(\ {}_i,\ {}_i) \propto\ {}_i^{-1} I(\ {}_i > 0) I(0 <\ {}_i < 1), \tag{1}$$

where $I(A)$ stands for the indicator function of the set A. Over the paper, we will refer to this model as the *Normal-non-hierarchichal model* (NN model). Details of the estimation procedure can be consulted in [9].

2.2 NBM Model Background

The NBM model idea came from the well-known multilevel regression and poststratification model (see, for example, [11]). In this model, they (1) divide the population into cells based on combinations of different attributes, (2) use the sample to obtain posterior simulations of the parameters, and (3) estimate the response variable within each cell, weighting then such estimates to estimate at the population level.

In the NBM model, we perform steps (1) and (2). However, in step (3), we use a model-based approach (see [7]), simulating the posterior distribution of the total number of votes per candidate for each of the polling stations in the state. Hence, the NBM model is used to predict the non-sampled values of the population. This gives an immediate estimate at the population level. The proposal is very flexible because there is an individual model per polling station. Nevertheless, it implies a major computational effort, since the number of simulated polling stations is large.

To perform step (1), we fit a multilevel regression for each polling station based on certain covariates. For each candidate, we model the number of votes at the polling station i, y_i, with a probability distribution $p(y|\ {}_1,\ {}_2)$, where, for a part of the parameters, say ${}_1$, we fit a multilevel regression based on certain predictors.

In the first multilevel model we tried, p was a truncated normal distribution. The posterior predictive check presented in Fig. 1 allows us to confirm that this distribution is not rich enough to capture the relevant structure of the true data generating process. This is because of the longer tails of observed data compared to the normal distribution, in particular with smaller candidates, which tend to have districts where they are extremely popular compared to the rest of the state.

We need a distribution that is also bell-shaped, but with heavier tails than the Normal distribution. A natural choice is the t-distribution. However, we also need to catch the high number of zero votes in some polling stations. This led us to the *heavy-tailed multilevel mixture model* (denoted as heavy-MM model) that was used on the 2018 quick count (see [1]). In this model, p is a mixture probability distribution.

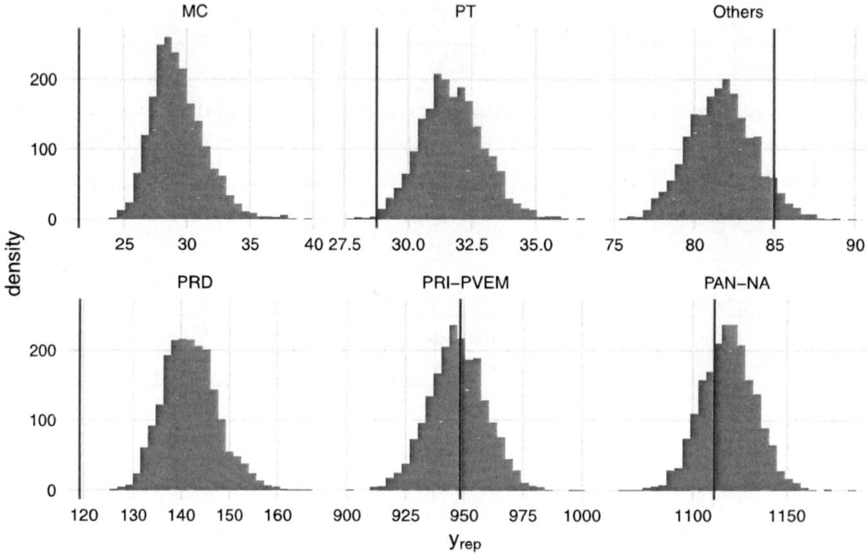

Fig. 1 Posterior predictive distributions of the number of total votes in the 2012 Governor's elections of Guanajuato using the multilevel model with a normal probability distribution. The red line indicates the total number of votes observed

$$p = q_k \; {}_0 + (1 - q_k)g,$$

where q_k is a mixing parameter, ${}_0$ is a degenerate distribution with mass at zero, and g is a t-distribution.

The most appealing part of this structure of multilevel models is that, in the absence of a response, it attracts the parameters of a group toward the group mean. As a result, we have a consistent treatment of missing data in samples and better interval coverage properties when the sample data is biased, compared to the combined ratio estimator and the NN model. However, an important disadvantage of the heavy-MM model is that the estimation method is slow. Running in parallel a candidate in each core, the heavy-MM model estimates with the complete sample in approximately 10 min, while, for instance, the combined ratio estimator runs in less than 30 s. On the election day, we have to produce estimates every 5 min. Hence, efficiency is extremely important.

One possibility used in 2018 is to do data split estimates independently for different geographical regions, parallelizing more the fitting. However, it is not desirable from the modeling point of view, as we cannot do partial pooling across regions or model covariance structures among candidates.

The NBM model, used in the 2021 elections, has the same structure as the previously described multilevel models, where p is a Negative Binomial distribution. The Negative Binomial distribution is shown to fit the data well, and it has one less parameter than the t-distribution. Therefore, the estimation procedure is faster. Having a faster estimation method allowed us to model the proportions of people who voted for each of the candidates jointly. In the Heavy-MM model, such proportions were modeled with independent distributions for efficiency reasons, even though it was a strong assumption.

In addition, there are other improvements that make the 2021 estimates better than those used in 2018. We have the use of Stan (see [14]). Stan code is readable, which makes the work more transparent. Stan also allows a number of diagnostics, which helped to obtain a faster reliable inference. Finally, a significant refinement was the adjustments we perform due to bias, which are discussed in the next section.

2.3 *The Bias Problem*

In 2018, the committee in charge of the quick count presented results for the election of president at 10:30 p.m. with 67.5% of the sample. Two of four candidates fell outside the confidence intervals given by the committee once all votes were counted. More detailed analyses showed that this happened because the sample was biased. For instance, in [10], the author contrasts the estimates obtained with the observed partial sample of 10:30 p.m. with the estimates obtained with random sub-samples of 67.5% of the selected sample. The difference in the estimates confirms that, within the observed partial sample, some polling stations were systematically more likely to be selected than others.

On this occasion, we conducted an exhaustive analysis of bias in previous elections and we faced the challenge of mitigating this bias at all stages of our workflow. In order to carry out this analysis, we needed a procedure to generate partial samples that resembles the true data generating process. Thus, we adjusted a survival model for the arrival times of the polling stations in the 2018 elections (see Fig. 2). The survival model details can be consulted in the Appendix.

We use the survival model to get a measure of the bias by state at different times. Simulating samples with arrival times and then censoring the samples, we compute the bias as the difference between the expected value of the estimates of the censored samples and the value observed with the full sample. Having a measure of the bias, we can adjust the estimation models accordingly.

The survival model has other uses. When calibrating the models, we can evaluate the models with partial samples, taking different levels of censorship to understand how far the real coverage of intervals can be from their nominal coverage. As a result, it allows to produce a set of rules to decide when to present results on the night of the election. Also, when designing the sample, we considered stratifications that mitigate bias under partial samples, using variables associated with the arrival time to stratify.

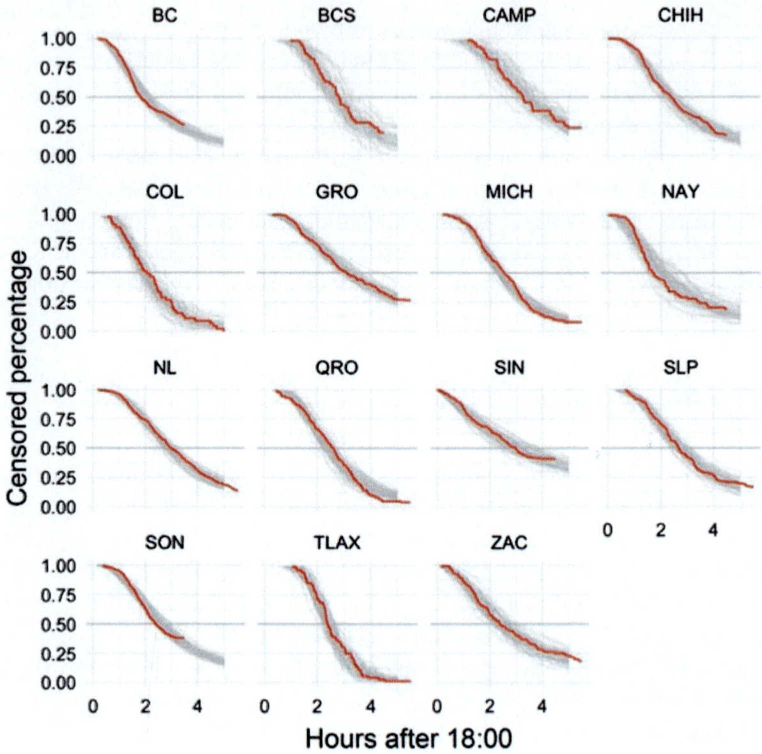

Fig. 2 The red lines are the observed data of the 2018 presidential elections. The gray lines show 50 simulations of the polling stations arrival time model

3 The NBM Model

In this section, we define the NBM model, which was obtained after several iterations of the Bayesian workflow in [12]. In addition to the model specification, we find it important to present three parts of the workflow that motivate several adjustments: (i) The consistency of our modeling assumptions, (ii) the fitting procedure, and (iii) the model adequacy.

3.1 Specification

We denote with NB the negative binomial distribution with a parametrization directly using a mean parameter, $_i \in \mathbb{R}^+$, and a parameter $_i \in \mathbb{R}^+$. For $y \in \mathbb{N}$,

$$\mathsf{NB}(y \mid {}_i, {}_i) \propto \left(\frac{{}_i}{{}_i + {}_i}\right)^y .$$

Our goal is to predict the participation, which is the share of people in the nominal list who voted, and the proportion of votes in favor of each one of the candidates. First, to compute the participation, we predict the total number of votes.

Let y_i be the total number of votes in polling station i, we assume that

$$y_i \sim \mathsf{NB}\left({}_i := n_i w_i, \ {}_i := \frac{n_i w_i}{{}^{part}_{s(i)}} \right),$$

where n_i is the size of the nominal list in polling station i, $w_i \in [0, 1]$ represents the proportion of people on the nominal list who voted in polling station i, and ${}^{part}_{s(i)} > 0$ is a parameter that depends on $s(i)$, the stratum to which the polling station i belongs.

The mean of y_i is given by ${}_i = n_i w_i$ and the variance is given by

$$\mathsf{var}(y_i) = {}_i + \frac{{}_i^2}{{}_i} = n_i w_i \left(1 + {}^{part}_{s(i)}\right).$$

Therefore, ${}^{part}_{s(i)}$ controls overdispersion relative to the mean.

We fit a multilevel-inverse-logistic regression model for the parameter w_i.

$$w_i = \mathsf{invlogit}\left({}^{part,0}_{s(i)} + {}^{T}_{part} x_i \right),$$

where x_i is the i-th row of X, a scaled matrix of covariates.

Now, we model the number of votes obtained for candidate k in polling station i as

$$y_{i,k} \sim \mathsf{NB}\left({}_{i,k} := n_i w_i \, {}_{i,k}, \ {}_{i,k} := \frac{n_i w_i \, {}_{i,k}}{{}_{s(i)}} \right).$$

Here, ${}_{i,k}$ represents the proportion of people who voted for candidate k out of the total number of votes in polling station i. Analogously to the participation case, ${}_{s(i)} > 0$ controls overdispersion relative to the mean, ${}_{i,k} = n_i w_i \, {}_{i,k}$, since the variance is given by

$$\mathsf{var}(y_{i,k}) = n_i w_i \, {}_{i,k} \left(1 + {}_{s(i)}\right).$$

Let p be the total number of candidates, the vector ${}_i = ({}_{i,1}, {}_{i,2}, \ldots, {}_{i,p})$ is given by

$${}_i = \mathsf{softmax}\left({}^{0,1}_{s(i)} + {}^{T}_{1} x_i, \ldots, {}^{0,p}_{s(i)} + {}^{T}_{p} x_i \right),$$

where

$$\text{softmax}(z_1, ..., z_p)_j = \frac{\exp(z_j)}{\sum_{i=1}^{p} \exp(z_i)}.$$

We consider the coefficients $_{part}, \ _1, ..., \ _p$ fixed over the entire state with a zero-centered Gaussian initial distribution and with a standard deviation of 0.25. However,

$$^{part,0}_{s(i)} \sim \text{N}(^{part}_{0}, \ ^{part}),$$
$$^{part}_{0} \sim \text{N}(-2, 1), \quad ^{part} \sim \text{N}^{+}(0, 0.5),$$

and, analogously, for $k = 1, ..., p$,

$$^{0}_{s(i)} \sim \text{N}(_{0,k}, \ _k),$$
$$_{0,k} \sim \text{N}(-2, 1), \quad _k \sim \text{N}^{+}(0, 0.5),$$

adding a hierarchical structure depending on the stratum. Having predictors at the stratum level reduces the unexplained variation within each stratum and, as a consequence, reduces the total variation, producing more accurate estimates. Finally, the parameters $^{part}_{s(i)}, \ _{s(i)}$ have a $\text{Gamma}(1, 0.1)$ initial distribution.

The priors influence the posterior inference with partial samples. Nevertheless, as data arrive, there is a moment in which the likelihoods will dominate the posterior inference. To justify the specific choices, the consistency of all the priors is analyzed in the following subsection.

On the day of the election, we used one covariate, a marginalization index. The index was created with the first principal component of the variables of housing equipment and services found in [6]. The decision was made by testing different combinations and balancing the time it took for the model to run. The marginalization index gave a significant improvement, while none of the other variables did. However, future work should certainly include exploring more covariates, interactions, or nonlinear effects in more detail.

3.2 Consistency of Our Modeling Assumptions

Before fitting the model to the data, it is useful to check that what we have encoded into the model is indeed coherent with our expectations. We can check if our priors are sound and correspond to domain knowledge by computing the a priori probability. In this case, we are interested in the proportion of votes per candidate. Hence, we can sample from the a priori distribution of the proportion of votes on the nominal list.

Figure 3 shows the proportion of votes for candidate 1 on the x-axis and the proportion of votes for candidate 2 on the y-axis. The choice of candidates 1 and 2 is arbitrary since the plots are analogous for any candidate. We can note that the priors

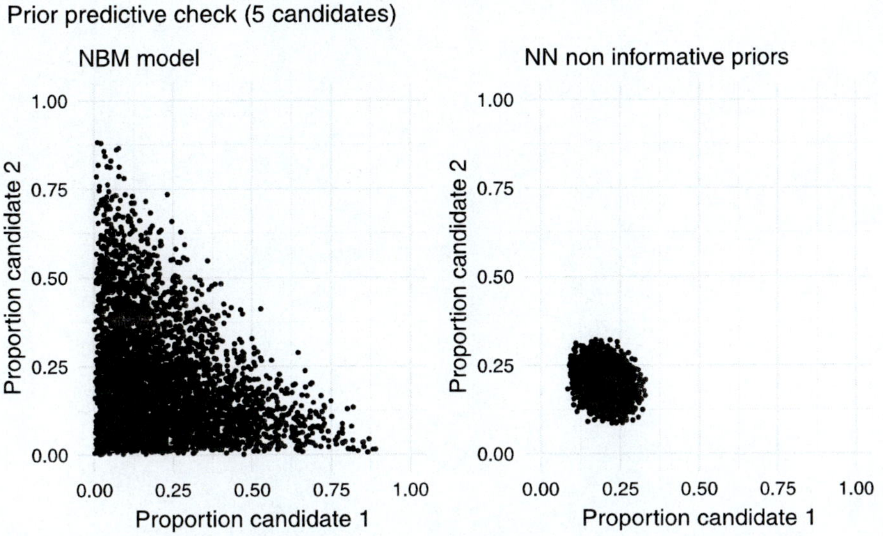

Fig. 3 The left block are samples from the a priori distribution of the proportion of votes with the NBM model in Nayarit, while the right block are samples from the a priori distribution of the proportion of votes with the NN model

of the NBM model provide a wide range of feasible values, while the prior of the NN model excludes data sets that could plausibly be observed. This exemplifies how some widely used prior distributions, recommended as being non-informative, do not actually respect the contextual knowledge. In the NBM model, most of the mass is between 0 and 0.5, but we still allow more extreme values, which is a reasonable configuration of the data.

Also, there are two characteristics of the posterior distribution that can identify a variety of possible problems, the z-score and the contraction (see [12] for definitions). Figure 4 shows the posterior z-scores as a function of posterior contraction. There are four potential outcomes here. (1) High posterior contraction with large (positive or negative) posterior z-scores reflects situations of overfitting to noise in the data. (2) Low posterior contraction with small z-scores reflects a poorly identified model. (3) Low contraction with large (positive or negative) z-scores indicates a substantial conflict between the prior and the likelihood. (4) High posterior contraction and low posterior z-scores, which is the case, reflect an ideal situation of good model fit.

The prior predictive checks show that the model is not excessively constrained by the priors and can still fit a wide variety of situations and data. Hence, we conclude that our choice of prior distributions is adequate.

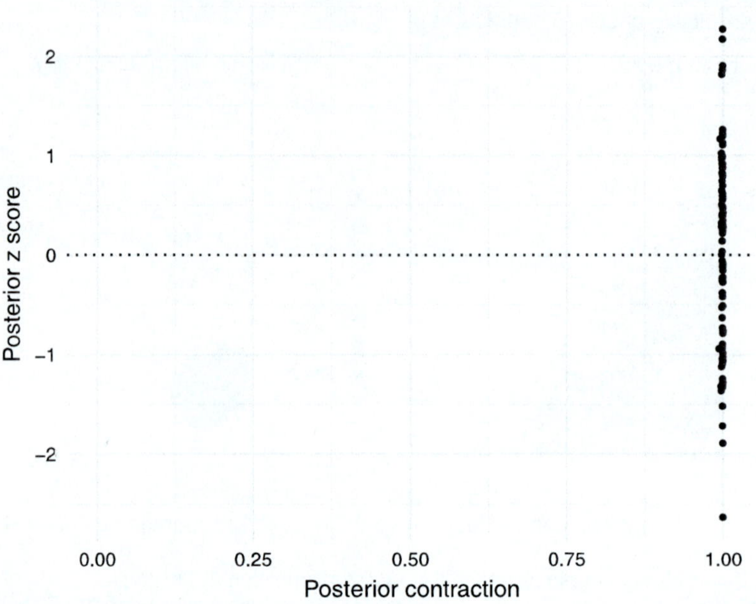

Fig. 4 Posterior contraction against posterior z-score of the number of votes per candidate. The posterior contraction quantifies how informative are observations relative to the prior model. The posterior z-score quantifies how accurately the posterior recovers the true model configuration (see [12])

3.3 Fitting Procedure

Since the model is written in Stan code, the fitting procedure for the data is straightforward from R using the package CmdStanR [13]. The implementation can be consulted and fully reproduced in the R package quickcountmx [5].

Once we obtain posterior simulations of the parameters, we use these to simulate the posterior of all polling stations in the sampling frame. First, for every polling station not in the sample, we simulate its vote counts according to the model. Then, we aggregate observed values for polling stations with simulated ones and obtain simulated vote counts for the total of polling stations. Finally, we summarize the aggregated samples to produce vote proportions and the corresponding intervals.

Stan provides diagnostic tools to evaluate the reliability of the inference. Passing all the diagnoses implied reparameterizing the model and using stronger priors. Furthermore, we can assess the faithfulness of our computational methods, before using the data, by applying them to posterior density functions arising from a variety of possible data. Figure 5 compares the posterior quantiles of simulated data with those of an independent uniform distribution (see [2]). We expect them to be close to the straight line $y = x$, which implies that we can recover the parameters that generated the simulated population.

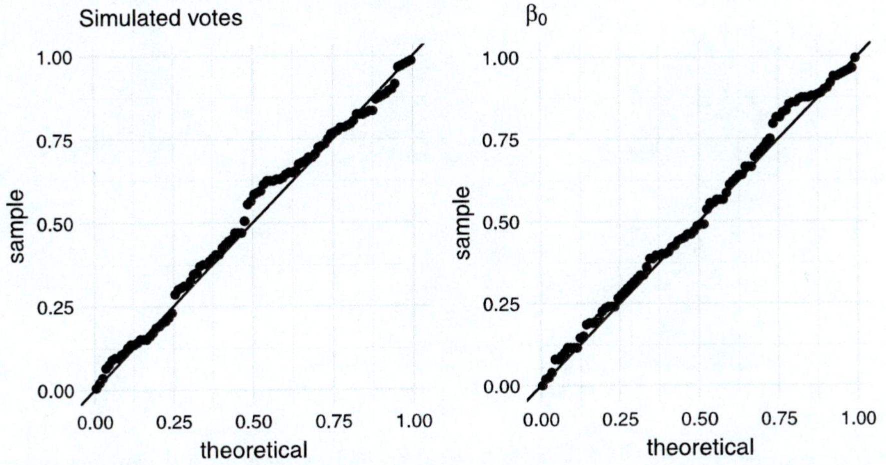

Fig. 5 The x-axis shows the quantiles of a theoretical uniform distribution. The y-axis shows the posterior quantiles (with respect to the posterior sample) of the true values of the total number of votes (left panel) and a fixed parameter that we denote with $_0$ (right panel). When the computational method is correct, the two distributions being compared are identical, meaning the Q-Q plot follows the line $y = x$

3.4 *Model Adequacy*

Now that we trust our inference, the next step is to check that the model produces simulations that are consistent with the observed data. This is the idea behind posterior predictive checks. Figure 6 shows the posterior predictive distribution in Nayarit, confirming it is compatible with all possible observations.

Afterward, we did Bayesian calibration with the data of the national elections of 2018. Bayesian calibration consists in checking that the model yields inferences that are well calibrated in a frequentist sense; specifically, we test that they provide posterior credibility intervals with approximately 95% coverage in repeated sampling.

The results of the NBM model appear in the left panel of Fig. 7. We can see the model works for the complete sample. However, when we have a considerable proportion of missing polling stations, the sample design and the models' covariates are not sufficient to correct the biases. In the quick counts that have been carried out, a considerable proportion of the polling stations (from 10% to 50%) is censored by time. To correct the bias in those cases, we will do the following adjustment depending on the observed sample percentage, which we will denote as p_{obs}.

First, we estimate the probability distribution of the total number of votes for each candidate in any polling station of the state with the model described in Sect. 3.1. Then, when simulating, instead of using the vector $_i$, we use a disturbance, given by

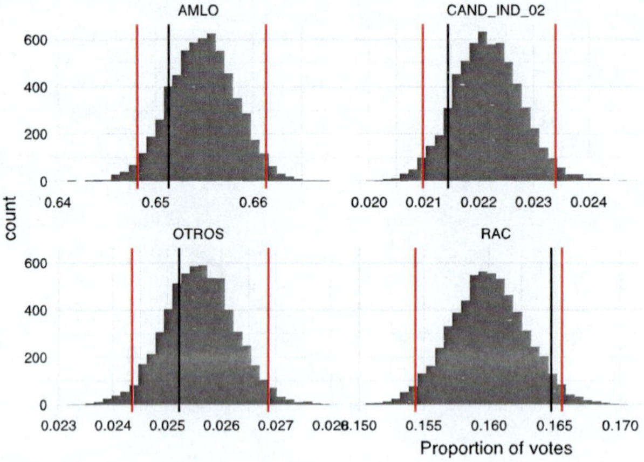

Fig. 6 Posterior predictive distributions of the number of total votes in the 2018 Governor's elections of Nayarit using the NBM model. The black line indicates the observed total number of votes. The red lines mark two standard deviations of the mean

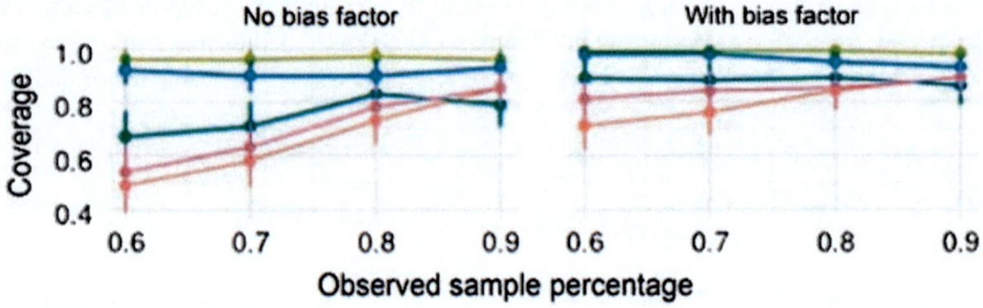

Fig. 7 Calibration of the NBM model for Nayarit. The points represent the percentage of times that the intervals cover the true value. The right panel shows the model with the bias adjustment described in Eq. 2, while the left panel shows the model with no bias adjustment. Each color represents a candidate

$$ {}'_{i} = softmax\left({}_{i,1} + {}_{i,1}, \ \ldots, \ {}_{i,p} + {}_{i,p} \right), \tag{2} $$

where ${}_{i,1}, , \ldots, {}_{i,p}$ are independent and identically distributed random variables whose initial distribution,

$$ \mathsf{N}(0, 0.1(1 - p_{obs})), $$

is set according to the bias observed in 2018 elections.

The size of the probability intervals produced grows as ${}_{i,k}$ grows and, as we approach the complete sample, ${}_{i,k}$ approaches 0 for any candidate k. Figure 8 shows an example of an estimation of the NBM model for a sample with and without the bias adjustment. We can notice the increase in the size of the intervals remains within reasonable ranges.

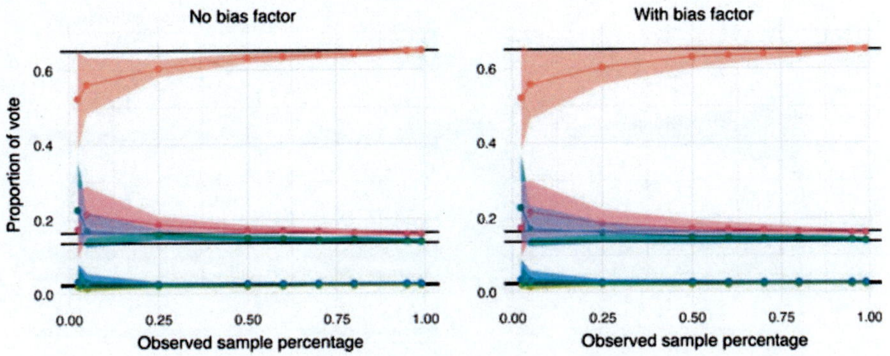

Fig. 8 Estimation with a fixed sample of the percentage of votes of the Nayarit gubernatorial election of 2018. The right panel shows the result with the bias adjustment described in Eq. 2, while the left panel shows the result with no bias adjustment. Each color represents a candidate

The results of the NBM model, with this perturbation, are in the right panel of Fig. 7. We can see the probability intervals coverage is much closer to 95% with censored samples. Hence, it provides us with a consistent way to adjust for bias and deal with missing data. Let us emphasize that, prior to these elections, statistical models for quick counts were tested with the full sample or with partial samples with random missing data. Figure 7 stresses the importance of having a method to test with partial samples with patterns close to the true data generating process.

To finish the validation of the model, we ran the calibration for the 2015 Mexican election data. Analyzing other elections would definitely contribute to a more complete picture of the quality of the model. However, the election processes have changed over time. In particular, in 2014, electoral reforms were introduced that markedly changed the functioning of elections in Mexico.

4 The 2021 Mexican Elections

4.1 Data and Sample Design

The committee explored several alternatives for the sample design. The chosen design was stratified with allocation proportional to the size of the strata. To define the strata and the sample size, we created different stratifications defined by combinations of variables. The variables we considered were those that are relevant in the survival model, since their use contributes to mitigating the bias. The details of this analysis and the final choices for each state are documented in [3, pages 128–134]. For Nayarit, we chose a sample with 10 strata and 20% of the total number of polling stations, which resulted in 334 polling stations.

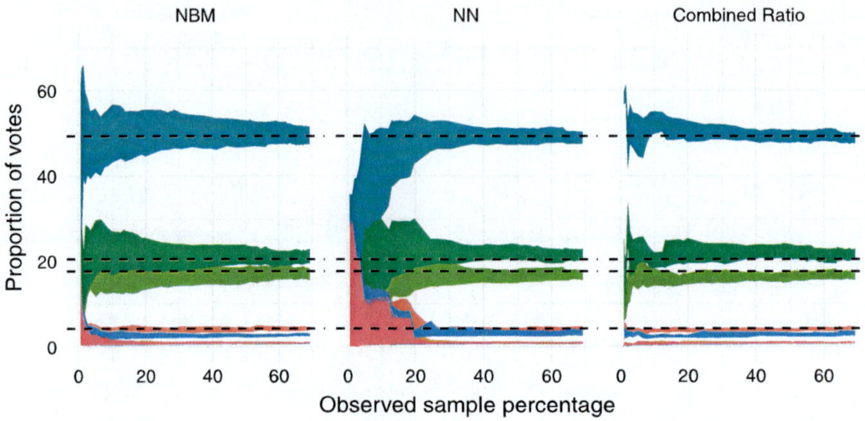

Fig. 9 Estimation of the percentage of votes of the Nayarit Governor's election of 2021 as the partial samples arrived (intervals of 95% of probability/confidence). The dashed lines are the true percentages and each color represents a candidate

4.2 *Results*

In the Nayarit Governor's election of 2021, the probability intervals of 23:35 were reported, with 220 polling stations in the partial sample. Figure 9 shows the monitoring of the intervals as the partial samples arrived, computed with the three methods in Sect. 2.1. Note the consistency provided by the proposed model, the model learns quickly from the data, so the trajectories narrow in a short time. Also, the intervals contain the true values all the time. Furthermore, we do not see a positive or negative slope in the trajectories, as is the case of the ratio estimator. Instead, the length of the intervals decreases as the sample arrives.

To illustrate the generality of the model, Fig. 10 shows the monitoring of the intervals as the partial samples arrived for Chihuahua. In Chihuahua, we reported the probability intervals of 00:25, with 225 polling stations out of 317.

5 Conclusions and Future Work

We presented the NBM model as an alternative model for the quick counts in the Mexican elections. In this context, we have an incomplete and biased sample. The NBM model provides (1) a consistent treatment of missing data, (2) better interval coverages, and (3) robustness, as it can fit well big and small candidates and provides a layer of protection when there are database errors.

We validated carefully the consistency of our modeling assumptions. Also, we presented a fitting procedure fast enough to be useful and a bias adjustment that results in probability intervals with coverage close to 95% with partial samples with certain types of bias.

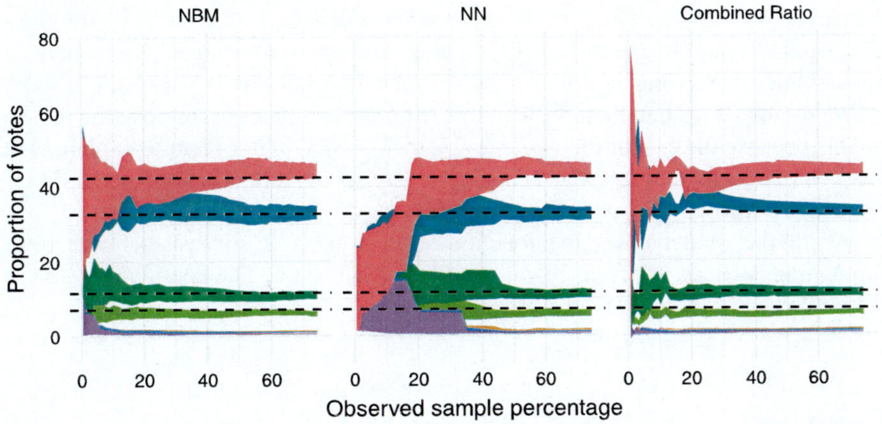

Fig. 10 Estimation of the percentage of votes of the Chihuahua Governor's election of 2021 as the partial samples arrived (intervals of 95% of probability/confidence). The dashed lines are the true percentages and each color represents a candidate

We discussed a survival model for the arrival times of the polling stations, used to generate partial samples with biases similar to those observed in previous elections. There is a lot to be done in that direction. More complex models need to be explored further. Such models could suggest the creation of new covariates to use in the negative binomial regression of the NBM model.

Regarding the NBM model, future work includes attempting a larger set of predictors, exploring more variables, and using interactions or nonlinear effects. The ultimate goal would be to have a set of covariates such that the model mitigates the bias without having to make perturbations when simulating.

Appendix

Polling Stations Arrival Time Model

We need a procedure to generate partial samples with patterns approximated to those seen on election days. Therefore, we model the arrival time of the polling stations to the database that the committee analyzes.

We use parametric survival models, censored by the right for the polling stations that did not arrive before 00:00 hrs. These models are stratified by state and we assume a log-logistic distribution for the arrival times. The data is the sample selected by the committee in 2018 and the sample observed at 00:00 the day after the election.

Most of the covariates used in the model come from databases provided by the electoral authority: the State, the type of section (urban or rural), the type of polling station (basic, contiguous, special, or extraordinary), the size of the nominal list, and the percentage of votes obtained for the three leading candidates of 2018. Additionally, we use a marginalization index, created with the first principal component of the variables of housing equipment and services found in [6].

We use the votes obtained for the three leading candidates to improve the fit. We do this because the rest of the variables by themselves do not seem to explain well the pattern of arrival times. This is not a problem, as the model is not being used for predictive purposes. However, more variables need to be explored further.

We fit the model using the function survreg from R survival package [15]. Simulating, we can verify in Fig. 2 the fit for each state. The coefficients and all the implementation can be reviewed in [4].

References

1. Anzarut, M., González, L.F., Ortiz, M.T.: A heavy-tailed multilevel mixture model for the quick count in the mexican elections of 2018. In: National Statistics Forum (FNE) and Latin-American Congress of Statistical Societies (CLATSE), pp. 1–13. Springer (2018)
2. Cook, S.R., Gelman, A., Rubin, D.B.: Validation of software for bayesian models using posterior quantiles. J. Comput. Graph. Stat. **15**(3), 675–692 (2006)
3. COTECORA 2020–2021: Criterios científicos, logísticos y operativos para la realización de los Conteos Rápidos y protocolo para la selección de las muestras (2021). Available from: https://repositoriodocumental.ine.mx/xmlui/bitstream/handle/123456789/119626/CGor202104-28-ap-15-Anexo.pdf
4. González, F., Meza, I., Ortiz, M.T., Anzarut, M.: cotecora-team-2/analisis-llegada. (v1.1). Zenodo (2021). https://doi.org/10.5281/zenodo.5703706
5. González, F., Meza, I., Ortiz, M.T., Anzarut, M.: cotecora-team-2/quickcountmx: Fuctions for the mexican quick-counts 2021. (v1.1). Zenodo (2021). https://doi.org/10.5281/zenodo.5546550
6. INEGI: Estadísticas Censales a Escalas Geoelectorales (2010). https://gaia.inegi.org.mx/geoelectoral/
7. Little, R.J.: Calibrated Bayes, an alternative inferential paradigm for official statistics. J. Off. Stat. **28**(3), 309 (2012)
8. Lohr, S.L.: Sampling: Design and Analysis. Chapman and Hall/CRC (2019)
9. Mendoza, M., Nieto-Barajas, L.E.: Quick counts in the Mexican presidential elections: A Bayesian approach. Electoral Stud. **43**, 124–132 (2016)
10. Orantes, G.: Conteo Rápido Para La Elección Presidencial de 2018 En México. Instituto Tecnológico Autónomo de México, Thesis (2019)
11. Park, D.K., Gelman, A., Bafumi, J.: Bayesian multilevel estimation with poststratification: state-level estimates from national polls. Polit. Anal. **12**(4), 375–385 (2004)
12. Schad, D., Betancourt, M., Vasishth, S.: Toward a principled bayesian workflow in cognitive science. Psychol. Methods **26** (2020). https://doi.org/10.1037/met0000275
13. Stan Development Team: CmdStanR: the R interface to CmdStan (2021). https://mc-stan.org/r-packages/
14. Stan Development Team: Stan modeling language users guide and reference manual, version 2.27 (2021). Available from: https://mc-stan.org
15. Therneau, T.M.: Survival-package: A Package for Survival Analysis in R (version 3.2-7, 2020). https://CRAN.R-project.org/package=survival

Assessing Hospitalization for SARS-CoV-2 Confirmed Cases by a Cross-Entropy Weighted Ensemble Classifier

Roberto Bárcenas and Ruth Fuentes-García

1 Introduction

In December 2019, a new coronavirus emerged as the cause of some cases of pneumonia in Wuhan, China. With the disease having spread in that country and into many other places around the world, the COVID-19 pandemic caused by SARS-CoV-2 is now considered a global public health emergency [1]. This virus has a great capacity to spread from person to person and to evolve into severe forms requiring hospitalization and, in the worst-case scenario, leading to death. Since it has only recently emerged, health systems have found it necessary to implement, as quickly as possible, standardized systems to maintain an adequate referral judgment for the early detection of patients at risk of evolving into a serious clinical situation. Assessing the human and technical resource capacities of hospital centers is also necessary to enable them to provide care when needed [2].

Since the onset of the pandemic, the double threat of this emergency has been pointed out. On the one hand, facing a new type of coronavirus for which, at first, there was insufficient knowledge to understand the risk at hand and its consequences. Therefore, vaccines have recently been developed to counteract the pandemic. The information to date, reported in [3], shows that most people infected with COVID-19 are mostly asymptomatic or experience mild symptoms and recover, not requiring hospitalization. These patients report fever (90% of cases), overall unwellness, dry cough (80%), chest pain (20%), and shortness of breath (15%). Chest radiographs with bilateral radio opacities and blood biometrics with the presence of leukopenia

R. Bárcenas (✉) · R. Fuentes-García
Departamento de Matemáticas, Facultad de Ciencias, Universidad Nacional Autónoma de México, 04510 Mexico City, Mexico
e-mail: rbarcenas@ciencias.unam.mx

R. Fuentes-García
e-mail: rfuentes@ciencias.unam.mx

I. Antoniano-Villalobos et al. (eds.), *Interdisciplinary Statistics in Mexico*,
Springer Proceedings in Mathematics & Statistics 397,
https://doi.org/10.1007/978-3-031-12778-6_4

and lymphopenia. Still, some of them required intensive care, although most were stable [4].

Second, the type of treatment patients receive in health units depends on how severe the disease is, some patients need very little intervention. Vital signs and oxygen levels will be monitored and fluids will be given to keep patients hydrated. When oxygen levels are low, they can be increased with a nasal cannula (a thin tube with two prongs that is placed under the nose and oxygen is administered through it) or an oxygen mask (similar to those worn by people with sleep apnea).

Because there is no definitive treatment to help patients recover from the virus, attention provided to patients in the COVID-19 special units comprises monitoring symptoms to determine if they are worsening and being alert for new symptoms that are very alarming. Most of the problems that warrant critical care are a severe respiratory or cardiovascular failure. When patients begin to need higher levels of blood pressure support, or if breathing is insufficient and they need more aggressive oxygen therapy, they should be admitted to the critical care unit.

In Mexico, during the early stages of the pandemic, COVID-19 clinical manifestation appears to be varied, ranging from asymptomatic or mild cases to situations as severe as acute respiratory failure, associated with consistent signs of pneumonia. Because of the diversity of respiratory symptoms in respiratory tract infections, the diagnosis for COVID-19 is essentially clinical and is characterized by fever and dry cough in conjunction with the criteria of the operational definition [5]. In the latter, health authorities consider suspected cases of viral respiratory disease persons of any age who, in the last 10 days, have shown at least one of the following major symptoms: cough, dyspnea (of seriousness), fever, or headache. Together with at least one of the following minor symptoms: myalgias, arthralgias, odynophagia, shivers, chest pain, rhinorrhea, polypnea, anosmia, dysgeusia, and conjunctivitis. The diagnosis of this viral infection comprises three principles: i) The epidemiological study, ii) Clinical manifestations, and iii) The diagnostic complement by a Reverse transcriptase-polymerase chain reaction (RT-PCR) test. Another relevant point is to record the day of symptom onset [6]. Any person satisfying the operational definition of a suspected case and diagnosed with a positive RT-PCR or antigen test by the National Network of Public Health Laboratories endorsed by the Institute of Epidemiological Diagnosis and Reference (INDRE) is declared a confirmed case.

Statistical techniques, including Machine Learning (ML), have proven to be helpful in addressing the COVID-19 pandemic. Data from SARS-CoV-2 positive patients has been integrated and analyzed using advanced algorithms to better understand the pattern of viral spread, to improve the speed and accuracy of diagnosis, to develop novel and effective clinical care treatment approaches, and to identify those most susceptible to a clinical picture of risk [7]. In addition, novel initiatives are using them for an outbreak and flare-up prediction, the monitoring of virus spread, the diagnosis and treatment, and for drug discovery and vaccine design. Medical communities around the world are turning their interest to these statistical approaches to help ease the stress on the hospital system. Because of the increasing number of patients requiring hospital beds, healthcare facilities in many countries eventually collapsed. Hence, medical decision support systems have been developed using ML tools. These are intended to help medical staff classify patient hazards, something that can be very critical, especially in scarce and very limited available resources. Thus, in many circumstances, they have been successful in determining the prospect of being infected or even predicting whether a person is a high-risk patient and allocating an ICU bed for survival [8]. However, challenges such as the quality of the training data and ML model limitations should always be kept in mind to adequately develop tools that can support the clinical users [9, 10].

This article proposes a learning model based on a set of classifiers from COVID-19 confirmed patient data in Mexico City, Mexico. Using a Random Forest to classify the type of patient (outpatient or hospitalized), we consider a measure of variable importance based on the Gini index for selecting the top-rated features. For each of these, a "weak" classifier is fitted, next we get the cross-entropy for the distribution of the fit predictions and the true labels. These values are used to determine a (normalized) set of weights pulling each classifier into an ensemble of rules. The key idea is that weak classifiers with a cross-entropy close to zero approach the best performance; therefore, we assign them a higher weight. As they move away from zero, their performance deteriorates and the weight in the classifier ensemble will be decreased. The aim is to provide a tool for health system authorities and a guide for the population to predict whether a person should be susceptible to hospital admission. This will provide a useful system for prompt triage of patients and alert potential candidates prone to reach a critical stage according to their feature profile.

2 Material

Because the pandemic is an emerging and ongoing process, monitoring new information regarding the characteristics of the virus, mechanism of transmission, specific clinical features, and severity of the disease remains necessary, to evaluate the impact of SARS-CoV-2 and to guide prevention and control actions to mitigate its spread both now and in the future. In this way, hospitals' resources could be managed with greater efficiency. Therefore, in Mexico, the Ministry of Health, through the General Directorate of Epidemiology, has the respiratory disease surveillance system (SISVER), which depends on the National Epidemiological Surveillance System (SINAVE), to dictate the procedures for epidemiological surveillance that includes the identification of cases, follow-up of contacts, as well as to proceed in aspects of the collection, handling, and proper shipment of samples and the analytical control available for the confirmation of cases.

Its work involves detecting suspected cases of viral respiratory disease that meet the operational definition in any health unit (public or private), aided by the corresponding epidemiological and laboratory surveillance, care, prevention, and follow-up operations. Through well-defined mechanisms, we consider the characteristics of patients that can be directed to different scenarios. From patients with mild symptomatology to patients suffering from or identified as serious as soon as they make first contact at the Health Centers. Note that a patient in need of life support and mechanical ventilatory support is not the same as admission to the ICU for some other reason. Triage in this type of patient directly influences their outcome. The overall experience and the experience of the staff are a guide to establish, assess, and select patients to be cared for in the correct areas of the hospitals.

2.1 Dataset Description

For this study, according to official information reported since March 1, 2020 and updated to June 30, 2021, data were obtained from the National Epidemiological Surveillance System Database (SINAVE) for the follow-up of potential cases of COVID-19 in Mexico City, under the permission of the General Epidemiology Directorate of the Ministry of Health.[1] This is an open-access repository comprising all COVID-19 cases, including suspects, tested by RT-PCR for confirmation of SARS-CoV-2 and which is updated over time. The information available for these confirmed cases contains the record of their demographic characteristics, such as age, sex, nationality, state, and the municipality where the case was detected. Clinical information includes symptomatology, while, for each patient, healthcare information registers the comorbidities.

From health system data, the following are incorporated: time elapsed since the onset of symptoms, hospital admission, probable diagnosis, as well as treatment status (outpatient or hospitalized), information on the diagnosis of pneumonia, admission to the intensive care unit (ICU), and whether the patient required invasive mechanical ventilation. As a reference, we have a key variable that represents the state of progression of the disease in each patient. Six cases define this assessment: not severe, recovered, severe, decease, treatment, and monitoring.

In the next section, we describe some aspects of the dataset and summarize the covariates of interest, which are later used to identify hospitalization as a response variable in the supervised classification.

2.2 Data Analysis

In our analysis, we have considered solely positive cases certified and reported by the National Institute of Diagnosis and Epidemiological Reference (INDRE), whose last update date is September 22, 2021. In total, there is a record of 242 660 confirmed positive SARS-CoV-2 patients in Mexico City. The specific variables associated with hospitalization for COVID-19 disease were investigated and we explored the coherence of this decision with the red outcomes given by the assessment. The variables useful to distinguish the most severe cases are sex and age. We also considered the

[1] The analysis was performed in the statistical software R. The dataset is available at https://datos.cdmx.gob.mx/dataset/base-covid-sinave.

distribution of symptom onset days in connection with the dynamics of the disease. At this point, there is an informed intuition, sustained in the literature [11], about the risk factors that lead patients to experience a complex disease, a tragic outcome, or simply moderate clinical manifestations. In most cases, the patient is declared outpatient and is sent home for treatment or monitoring. However, the prevalence of the severe disease is higher in patients with obesity, diabetes, chronic renal insufficiency, COPD, immunosuppression, and hypertension, particularly from male patients exceeding 65 years of age.

The most affected age group, in severe scenarios, corresponds to patients over 50 years (81.2% of all severe cases), with a median mortality rate of 65 years. Also, the higher percentage of hospitalized patients are between 50 and 75 years of age (56.6% of the total number of hospitalizations). By gender, we found the same number of cases in women and men, but there are differences between hospitalized and outpatient cases. In women, there are 115 362 outpatients and 9 807 hospitalized, which, compared with 103 067 and 14 424, respectively, in men, produce a clear imbalance. Meaning, hospitalization in men is 47% higher than in women, whereas their outpatients are 12% lower compared to their female counterparts.

If we compare hospitalized versus outpatient patients, the first finding is that 90% of the positive cases were outpatient, whereas 10% required hospitalization. Regarding lethality, based on these data, it is estimated at 5% for Mexico City (including both outpatients and hospitalized patients). The upper plot on Fig. 1 shows the observed percentages when grouping each type of patient regarding their reported outcome. We find that low-risk cases (follow-up and monitoring) are outpatient, while more severe cases are hospitalized. In particular, 93% of outpatients corresponding to monitoring class, however, to maintain an adequate scale are omitted in the plot. That is, for a clearer visualization, the graph does not show the data bar for patients under home monitoring. With hospitalized patients, 4.9% of patients are being discharged, but, 4.6% of hospitalized patients die. Consequently, greater preventive vigilance is suggested, both in terms of information given to the population and in the procedures to promptly attend to each case in order to avoid this type of event. Knowing in advance the risk of having a person and updating forthwith as the condition evolves or is worsened by other factors that may suddenly appear is decisive.

Mexico City has a protocol with standardized concepts and procedures for conducting epidemiological vigilance. Following these surveillance parameters, the probable diagnosis of COVID-19 patients is categorized into two cases: Influenza Type Illness (ETI) and Severe Acute Respiratory Infection (IRAG). The main difference is that in IRAG cases, registration is made for all who meet the definition of a suspected case of Viral Respiratory Disease and who show any of the following symptoms: dyspnea, chest pain, or desaturation. The ETI classification is seen in outpatient cases; however, when it evolves with serious symptoms and signs, it is classified as IRAG and is referred to the hospitalization in health units with this service. The most common cases of acute respiratory infection with favorable evolution are found in outpatients, while serious cases are found in hospitals. As shown in Fig. 1, bottom left, most of the diagnoses with ETI are recorded as outpatient (83.6%), enabling us to distinguish a pattern of non-hospitalization under this label.

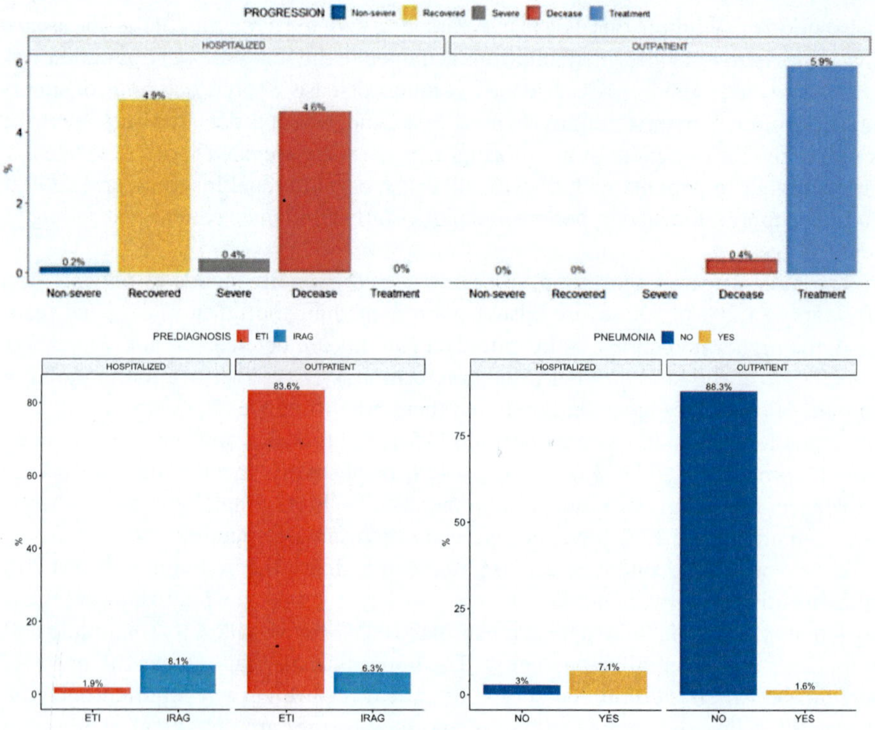

Fig. 1 Progression state, respiratory class infection system, and preliminary diagnosis of COVID-19 proportions, all separated by type of Patient

The opposite occurs with the preliminary diagnosis of IRAG, as it is presented in a close frequency between the two categories of patients, 8.1% for hospitalized ones and 6.3% for outpatients.

A clinical complication often encountered is pneumonia, as viral infections in the respiratory tract influence several of the host defense factors and pave the way for subsequent bacterial infection. In [12], it is stated that 15–20% of patients with COVID-19 experience moderate to severe forms of pneumonia. While it has been found that, some patients have other symptoms, such as myalgias, headache, sore throat, and diarrhea. The cases of pneumonia present with swelling on chest X-ray and opacity on chest CT scan [13]. Following Fig. 1 bottom right, we noticed that it is a differential factor between outpatient and hospitalized cases. In our example, we could recognize that detection of pneumonia occurs in 8.7% of all cases, with hospitalization required in 7.1%. Furthermore, for 88.3% of the cases in which pneumonia is not present, it is possible to assign them as outpatients. A challenging decision is made in the 3% of patients where pneumonia is not detected, yet are considered for hospitalization on other grounds. Therefore, it is desirable to have additional guidelines to accompany the assessment of patients with COVID-19 and to forecast scenarios in the presence of pneumonia.

3 Cross-entropy Weighted Ensemble Classifier

Ensemble learning, as a classification rule, is an approach that makes predictions based on the solution of several different classifiers. By combining multiple single predictions, the ensemble model tends to be more flexible (less bias) and less sensitive to the data (less variance), in contrast to other learning approaches. Bagging and boosting are the two most popular ensemble methods. Bagging involves training a group of individual models in parallel, where each uses a random subset of data and ranked with a voting consensus. In Boosting, the group of individual models is trained sequentially. Here, each single classifier learns from the errors made by the previous model and the output is a linear (weighted) combination of the classifiers [14].

The method we will consider is the algorithm proposed by Breiman, Random Forests (RF), which is a generalization of the Bootstrap Aggregating (Bagging) method [15]. RF builds multiple trees based on the CART algorithm [16] and then combines the predictions or classifications (regression or classification) given by each tree. The latter is intended to reduce the variance in the individual trees and thus improve the performance of the method. We used 39 covariates to train the algorithm, where the response variable corresponds to the type of patient. This means that we are dealing with a binary classification problem with labels: hospitalized or ambulatory. As in most classification schemes, training and test sets are employed. Also, a validation set of parameters is drawn for the initial tuning of some quantities to be used in the training. In this case, we opt for K-fold cross-validation (with $K = 10$), which randomly divides the data into K blocks of equal size. Each of the blocks is left out in turn and the other $K - 1$ blocks are used to train the model. The excluded block is predicted, using these predictions to compute a performance measure; in this application, we use the accuracy.

A fundamental aspect is that RF allow us to implicitly generate measures of importance for the variables in the model, which can be constructed using the Gini index or the random permutation of variables [17]. The goal of these measures is to quantify the impact that each predictor has on the response variable, therefore determining its relevance or predictive effect, guiding the formulation of variable selection algorithms [17]. Information gain is based on the idea of entropy, which is the degree of uncertainty or impurity. For classification, and in particular, within the CART approach, the Gini index measures the degree to which a given variable, chosen at random, classifies wrongly. Note that the Gini index and Entropy do not represent cost functions; however, they measure the impurities at each node used to divide the branches in RF. The Mean Squared Error (MSE) is the cost function used for regression, while the cross-entropy cost function is used for classification.

The cross-entropy is a measure of the difference between two probability distributions for a given random variable. In some sense, it determines the distance between two distributions and in ML is widely used as a loss function when optimizing classification models. Its minimization results in parameter optimization. To this end, the average of the cross-entropy between all the labels of the training samples and the corresponding predictions are usually obtained. Logistic regression (a linear classification algorithm) and artificial neural networks are examples of its application.

To determine the cross-entropy between the measures P and Q, we can calculate the entropy for Q using the probability weights of P. Formally, it is defined as follows:

$$H(P, Q) = \mathbb{E}_{x \sim P}\big[-\log Q(x)]. \tag{1}$$

However, since we do not know the true distributions of the actual labels, an approximation is used. Next, given the actual values of the target variable data as y_i, $i = 1, \ldots, n$, and the predictions $\hat{y}_i$; the cross-entropy is estimated by the following formula:

$$H(y, \hat{y}) = -\frac{1}{n}\sum_{i=1}^{n}\big[y_i \cdot \log(\hat{y}_i) + (1 - y_i) \cdot \log(1 - \hat{y}_i)\big]. \tag{2}$$

This is performed in the same way as in cross-entropy-based error reduction methodologies, such as neural networks. From the above, this proposal proceeds in two stages. In the first, we train a RF and identify the variables of maximum importance under the measure given by the Gini index. Next, we use these variables to construct an ensemble of (weak) classifiers.

Then, a stump[2] tree will be fitted for each variable selected as important in the RF classifier [14]. During the training, the cross-entropy between the distribution of true labels and the predictions will be calculated. Before the ensemble is settled, the weights of each of these classifiers are determined as a normalized measure of the cross-entropies obtained among the classifiers. This is done in order to ensure that the weights add up to one, hence the ensemble is well defined. Once the weights of each stump have been calculated, the cross-entropy-based ensemble is proposed as follows:

$$G(x) = \sum_{i=1}^{J} \alpha_i^* g_i(x), \tag{3}$$

where the quantities α_i^* are the normalized weights (such that $\sum \alpha_i = 1$) based on the cross-entropy and associated to the weak classifiers given by $g_i(x)$, and also, J denotes de number of stumps that results equal to the number of variables selected. Therefore, equation (3) provides the final classifier for the data sample x. From this classifier, the prediction for the test data is produced to validate the performance of this approach.

[2] A stump is a tree with a depth equal to one.

Notice that for simplicity and to combine the criteria, we used the Gini index as a criterion in the construction of the trees, in order to prune them and prevent them from being correlated in the first stage of the proposal. Then, for the ensemble, we considered cross-entropy, which is also closely related to implementing neural networks and has been observed to be useful when the goal is to improve the classification performance.

The above procedure is summarized in the following box:

Cross-entropy weighted classifier:

(1) Train a Random Forest with all the covariates.
(2) Determine the feature importance based on the information gain measure: Gini index.
(3) Select the most important variables.
(4) Fit a stump tree using each variable selected.
(5) For each weak classifier, calculate the cross-entropy between the distribution of true labels and the predictions.
(6) Normalize the quantities in (5) and define them as the weights of the classifier ensemble.
(7) Calculate the cross-entropy-based ensemble classifier by the formula 3.

4 Results

In this section, we compare the performance of a RF technique and novel approach based on an ensemble of classifiers derived from stump trees to identify hospitalized patients.

Using 169 863 instances corresponding to 70% of the total, the first stage of the methodology trains a RF model with the 39 features, computing the importance of the variables. Here, we have 152 712 outpatients and 17 151 hospitalized. According to its definition, it is the sum of the number of splits (of all trees) that includes the given feature, proportionally to the number of samples it divides. We can then obtain a quantification of the global importance of each feature via the Gini index. This measure corresponds to the sum over the decrease of the Gini index by the splits on a predictor averaged over all trees.

Figure 2 shows a graphical representation of feature importance. The 12 most important variables are presented in descending order, where we observe a reduction in the Gini index for each variable. The features with the largest mean decrease in Gini index are as follows: Pneumonia, Preliminary Diagnosis, Dyspnea, Age, Days elapsed to symptoms, Contact with someone infected, Suddenly Attack, Occupation, Fever, Polypnea, Diabetes, and Hypertension. Here, the feature Suddenly Attack involves an effect to the general state and expresses the patient presents a condition of discomfort that precipitated. It is associated with prostration plus a state of weakness

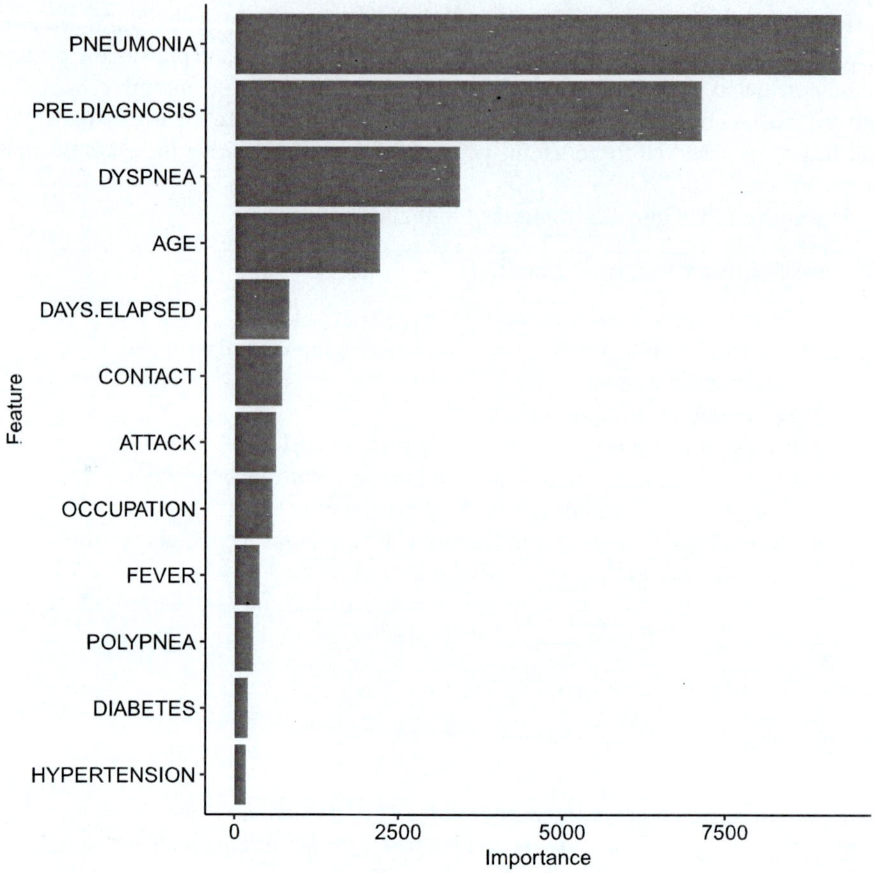

Fig. 2 Feature importance by the Random Forest training

and fatigue. Thus, it refers to any health problem that reflects a poor condition which turned up.

Three groups of variables are observed: the first, a group on diagnostic and sociodemographic characteristics; the second group is related to symptoms; and the third group concerns previous illnesses. In addition we noticed that the problem is reduced sensibly, going from 39 features to a sensibly lower number. The clinical variables labeled as Pneumonia and Preliminary diagnosis appear as the most important, as suggested by the statistical analysis in Sect. 2.2, followed by Dyspnea and Age.

Next, considering the eight most important features, we train the ensemble of weak classifiers based on cross-entropy. Once we have trained the algorithms, we compare the performance of the algorithms using the test sample. Table 1 shows the contrast of both models for the unused observations, removing any possible overfitting. Note that we are faced with an unbalanced problem, since, there are 7 350 inpatients and

Table 1 Performance comparison: random forest versus cross-entropy weighted Ensemble

Random Forest	True	
Predicted	**Hospitalized**	**Outpatient**
Hospitalized	4297	1818
Outpatient	3053	63629

Accuracy	**89.86%**
Sensitivity	58.46%
Specificity	97.22%
Precision	70.27%
F_1	63.82 %
Balance Acc	77.84%

Ensemble Proposal	True	
Predicted	**Hospitalized**	**Outpatient**
Hospitalized	6638	**5505**
Outpatient	**712**	59942

Accuracy	**91.46%**
Sensitivity	90.31%
Specificity	91.58%
Precision	54.66%
F_1	68.1%
Balance Acc	90.95%

65 447 outpatients of the 72 797 observations in the test set (30% of the total), and this influences the performance of the classifiers.

On the upper left side of Table 1, we display the confusion matrix for the RF with all the variables. We note a difficulty in correctly classifying the class of interest, since 4 297 hospitalized are well classified yet 3 053 are wrongly predicted as outpatients. In this case, we are interested in patients susceptible to hospitalization. On the other hand, as is often the case in unbalanced problems, the majority class is the one with the most correct classifications, we observe that there is no inconvenience in determining outpatients using the RF algorithm. We found 63 629 correctly classified outpatients and only 1 818 misclassified patients in this group. In the bottom right side of Table 1, the confusion matrix of our proposal is shown. It is interesting to see a significant improvement in the classification of the minority class, since now 6 638 of the total number of hospitalized patients are detected and there is a misclassification for 712. For the outpatient class, we note the greatest change in the increase of errors, reaching 5 505 predicted as hospitalized, which in fact would not be so. This aspect should be emphasized in the discussion.

On the upper right side of Table 1, we show the classification metrics obtained from the confusion matrix. For the RF, although the accuracy is 89.86%, we have that the sensitivity is low (58.46%), which affects the average F_1 score and Balance Accuracy measures, whose values are 63.82% and 77.84%, respectively. In a sketchy comparison, on the bottom right side of Table 1, we have that our proposal provides an accuracy of 91.46% (slightly higher than the RF), although we note that there is an improvement in at least three metrics. Sensitivity increases to 90.31%, while F_1 score and Balance accuracy reach 68.1% and 90.95%. Despite a decrease in specificity, which previously was 97.22% and in our approach 91.58%, we can say that despite trying to improve in sensitive cases requiring hospitalization, we achieved a positive result. One aspect to improve is the precision referred to predicted cases, where in our proposal it is reduced to 54.66%, derived from having many False Positive cases in the hospitalization class. This is addressed in the discussion and can be justified according to a specific target.

5 Discussion

First, we discuss the rationale for choosing an RF algorithm as the baseline. In this method, each tree in the ensemble is generated from a sample drawn with substitution (e.g., a bootstrap sample) from the training set. When splitting a node during tree construction, the split it chooses is the best split among a random set of features. Because of this randomness, as a rule, the bias of the forest increases while its variance decreases due to averaging, often more than compensate for the increase in bias, and therefore typically generates a better model.

Furthermore, RF have helped pinpoint informative predictor variables. This methodology generates measures of importance for the variables, which in turn allowed us to select the features that were truly relevant. Since ensemble learning is an approach that makes predictions based on a number of different models, by combining individual classifiers, we recover the theoretical aspects of PAC learning [18], turning weak classifiers, although informative in the sense of having relevant variables, into a strong classifier with a better performance. Given the nature of the ensemble and the fact that it considers cross-entropy, we highlight that this measure has proven important in different applications. In ML it is used when optimizing a logistic regression or a neural network model. Note also its interpretability; particularly when two distributions are identical, the cross-entropy between them is equal to the entropy of the probability distribution. This means that the cross-entropy of two distributions (actual and predicted) that have the same probability distribution for a class label is equal to 0. Therefore, the predicted class probabilities are identical to the probabilities of the training dataset, that is, zero loss.

In particular, applied to our proposal, we have that the choice of stumps as a weak classifier in the cross-entropy-based ensemble certainly works in that way. The latter, since when considering any of the selected features, a binary partition of the space is produced and it gives a naive classification, which when added with the others, leads us to the results we presented.

However, we note that there is a rare phenomenon, because in the allocation of importance in the RF, we found that, on average, the use of variables associated with comorbidities has relatively less importance than others. Then, the choice of these eight variables of larger importance obeys the fact that up to that number of variables, the weak classifiers are informative. If we introduce variables other than these eight, it leads to a naive classifier, because it does not learn and completely misclassifies all the observations. A novel idea would be to use random combinations of features in weak classifiers and form ensembles.

Regarding the results, classification methods must consider the imbalance and asymmetries in the predictions. Since in this case, the most important issue is to find candidates for hospitalization, given the covariates, improving the RF model is valued. From the computational perspective, once we have determined the variables of importance, our algorithm has the advantage of being able to be run in parallel, since the classifiers that integrate the ensemble can be executed independently and this, together with the use of parallel and distributed computing techniques, can improve performance time and scaling of any algorithm. Notice that, in theory, learning algorithms use results from Uniform Strong Laws stated in general spaces. The key result is the consistency of a wide range of techniques, e.g., RF and Boosting. We consider the accuracy as a reference to establish that, practically, our methodology performs as well as RF. However, the important consideration is to obtain proper metrics for the analysis of unbalanced data. Therefore, in the summary tables, we considered different metrics (sensitivity, specificity, and balance accuracy), which provide new information and allow us to quantify the decisions differently.

On the other hand, thinking in a conservative scenario, the health system must afford to care for all patients, which means that counting outpatients as candidates for hospitalization, i.e., misclassified outpatients, represents an affordable cost. The foregoing is proposed, intended only as a guideline for specialists, in the sense of focusing attention on these patients. This means that the response of the classifier does not imply that they are admitted immediately to the hospital, rather they should receive a closer follow-up, as well as guarantee the resources and facility for them to be examined in a health unit. The fact that a patient is classified as hospitalized by our method does not mean that they will be treated invasively and automatically.

The aim is to address and discuss the fact of detecting the greatest number of suspected cases for hospitalization, as a measure to support the authorities. This would answer, in part, the concern about the collection of information. It would alert and promote remaining vigilant to those patients referred home, but observed to be susceptible to hospitalization. This would follow them up and complete the filling out of the information. Therefore, we strongly believe that this type of initiative will promote the flow of information necessary to address questions about information distortions and imbalances in patient care and preserve all variables, since, at this stage of the research, we do not know if any of the variables will provide new information in the future.

Efforts are underway to provide more robust evidence to define groups at risk for severe complications associated with COVID-19 ([19–21]). The current evidence shows hints of predispositions in suspected and confirmed patients. There are suggestions that the risk of moderate to severe COVID-19 manifested as pneumonia is increased in adults over 65 years of age and males, with specific symptomatology and suffering from chronic conditions such as diabetes, hypertension, and cardiovascular disease. Patients with weakened immune systems are also more widely affected. These cases may require hospital care [22]. The criteria should focus on the immediate detection of suspected severe cases, to contain the progression to irreversible endings. To achieve this goal, it is necessary to establish highly sensitive systems that detect the potential severity, but also the feasibility of reaching it, according to the clinical profile and further information. The purpose of this is to trigger preventive surveillance actions and take the corresponding control measures aimed at reducing the risks of an adverse evolution. Aligning epidemiological surveillance with standardized measurements of accessible features to the population that can be entered into a system would make it possible to detect most serious cases through the most frequent signs and symptoms reported to date, as well as through the epidemiological records of the cases and information that is updated in real time. Finally, the important point is that it also serves as an alert for the patient, indicating that they should take the necessary care and that if respiratory or other complications appear, they should go immediately to the hospital for prompt attention.

6 Conclusion

Hospitalization of patients with acute respiratory COVID-19 infection caused by the SARS-CoV2 virus is a critical decision, as it has a direct effect on the potential outcome. In addition, it leads to an allocation of resources (bed, care, and medical personnel) that, given the pandemic, are limited. Based on the knowledge of a set of features, an ensemble of classifiers weighted by the cross-entropy information measure is introduced. The aim is to contribute toward the enhancement of a judgment and to assist the population in knowing the risk they are exposed to the disease.

In this work, we have presented a proposal that incorporates well-known methods aimed at solving the problem of prioritizing hospital care due to the COVID-19 pandemic. We start with a RF algorithm, using bagging as the ensemble method and decision tree as the individual model. Determining a set of important variables, based on an information measure (Gini index), has been key for the classification task. By proposing an ensemble of weak classifiers with each of the eight most important variables, we have achieved a significant improvement in the goal of capturing the label of our interest (hospitalization), without deteriorating the accuracy with the gain of improving other measures (F_1-score and Balance Accuracy) and the interpretability of the classification. The results show an improvement in the detection of cases susceptible to hospitalization, with an accuracy of 91.46%. In a restrictive scenario, there is a preventive alert to patients, even though under the established criteria should not be admitted, to remain under monitoring to anticipate the evolution of the disease to a severe stage.

To train the algorithm, we used different variables and responses as a reference to evaluate the severity of COVID-19 and to assess hospitalization. In the future, it is expected that this type of approach will work to assist in decision-making about the admission of the patients in most need, for efficient management of medical resources, which we know are limited, especially in a pandemic such as COVID-19. It also assists the population in the form of an automated tool that alerts them about the actual risk they face when they are diagnosed with the virus. This will bring them to contact their medical unit for timely care, and while it does not dictate instant hospitalization, it can be an early signal to monitor more carefully the course of these patients.

Finally, this analysis is relevant to the particular period at hand, that is the early stages of the pandemic. Therefore, the effects of variants of the virus and vaccination were not available at the time. They would eventually update the classification if the period of analysis is extended.

Acknowledgements This research has been partially supported by PAPIIT Project IN118720 and a postdoctoral grant, UNAM-DGAPA, Mexico. The authors are also thankful to the two anonymous reviewers whose comments helped improve the paper.

References

1. Coronavirus disease (COVID-19) [Internet], 16 Sept 2021. https://www.who.int/health-topics/coronavirus
2. World Health Organization: Coronavirus disease (COVID-19) technical guidance: Unity Studies: Early Investigation Protocols. Ginebra: WHO, 15 May 2020. https://www.who.int/emergencies/diseases/novel-coronavirus-2019/technical-guidance/early-investigations
3. World Health Organization: WHO living guideline: drugs to prevent COVID-19, interim guidance, 2 March 2021. World Health Organization (2021). https://apps.who.int/iris/handle/10665/339877
4. Estimated COVID-19 Burden [Internet], 6 July 2021. https://www.cdc.gov/coronavirus/2019-ncov/cases-updates/burden.html
5. Lineamiento estandarizado para la vigilancia epidemiológica y por laboratorio para la enfermedad respiratoria viral, August 2021, Secretaría de Salud México, Subsecretaría de Prevención y Promoción de la Salud Dirección General de Epidemiología. www.gob.mx/salud
6. Grupo de Trabajo Mexicano COVID-/COMMEC: Guía COVID-19 para la atención del paciente crítico con infección por SARS CoV 2 Colegio Mexicano de Medicina Crítica. Med Crit. **34**(1), 7–42 (2020). https://doi.org/10.35366/93279
7. Alimadadi, A., Aryal, S., Manandhar, I., Munroe, P.B., Joe, B., Cheng, X.: Artificial intelligence and machine learning to fight COVID-19. Physiol. Genom. **52**(4), 200–202 (2020). https://doi.org/10.1152/physiolgenomics.00029.2020
8. Peiffer-Smadja, N., Rawson, T.M., Ahmad, R., Buchard, A., Georgiou, P., Lescure, F.-X., Birgand, G., Holmes, A.H.: Machine learning for clinical decision support in infectious diseases: a narrative review of current applications. Clin. Microbiol. Infect. **26**(5), 584–595 (2020). ISSN 1198-743X. https://doi.org/10.1016/j.cmi.2019.09.009
9. Verdonk, C., Verdonk, F., Dreyfus, G.: How machine learning could be used in clinical practice during an epidemic. Crit. Care **24**(1), 265 (2020). https://doi.org/10.1186/s13054-020-02962-y
10. Roberts, M., Driggs, D., Thorpe, M., Gilbey, J., Yeung, M., Ursprung, S., Aviles-Rivero, A.I., et al.: Common pitfalls and recommendations for using machine learning to detect and prognosticate for COVID-19 using chest radiographs and CT scans. Nat. Mach. Intell. **3**, 199–217 (2021)
11. Bello-Chavolla, O.Y., Bahena-López, J.P., Antonio-Villa, N.E., Vargas-Vázquez, A., González-Díaz, A., Márquez-Salinas, A., Fermín-Martínez, C.A., Naveja, J.J., Aguilar-Salinas, C.A.: Predicting mortality due to SARS-CoV-2: a mechanistic score relating obesity and diabetes to COVID-19 outcomes in Mexico. J. Clin. Endocrinol. Metab. **105**(8), 2752–2761 (2020). https://doi.org/10.1210/clinem/dgaa346
12. Sánchez-Talanquer, M., et al.: Mexico's Response to COVID-19: A Case Study. Institute for Global Health Sciences (2020). https://globalhealthsciences.ucsf.edu/sites/globalhealthsciences.ucsf.edu/files/mexico-covid-19-case-study-english.pdf

13. Cummings, M.J., Baldwin, M.R., Abrams, D., Jacobson, S.D., Meyer, B.J., Balough, E.M., Aaron, J.G., Claassen, J., Rabbani, L.E., Hastie, J., Hochman, B.R., Salazar-Schicchi, J., Yip, N.H., Brodie, D., O'Donnel, M.R.: Epidemiology, clinical course, and outcomes of critically ill adults with COVID-19 in New York City: a prospective cohort study. The Lancet **395**, 1763–1770 (2020). https://doi.org/10.1016/S0140-6736(20)31189-2
14. Schapire, R.E., Freund, Y.: Boosting: Foundations and Algorithms. The MIT Press (2012)
15. Breiman, L.: Bagging predictors. Mach. Learn. **24**, 123–140 (1996). https://doi.org/10.1007/BF00058655
16. Breiman, L., Friedman, J.H., Olshen, R.A., Stone, C.J.: Classification and Regression Trees. Chapman & Hall (1993)
17. Breiman, L.: Random forests. Mach. Learn. **45**, 5–32 (2001). https://doi.org/10.1023/A:1010933404324
18. Valiant, L.G.: A theory of the learnable. Commun. ACM **27**(11), 1134–1142 (1984)
19. Jordan, R.E., Adab, P., Cheng, K.K.: Covid-19: risk factors for severe disease and death. BMJ **368**, 1198 (2020). https://doi.org/10.1136/bmj.m1198
20. Assaf, D., Gutman, Y., Neuman, Y., et al.: Utilization of machine-learning models to accurately predict the risk for critical COVID-19. Int. Emerg. Med. **15**, 1435–1443 (2020). https://doi.org/10.1007/s11739-020-02475-0
21. Antonio-Villa, N.E., Bello-Chavolla, O.Y., Vargas-Vázquez, A., Fermín-Martínez, C.A., Márquez-Salinas, A., Pisanty-Alatorre, J., Bahena-López, J.P.: Assessing the burden of coronavirus disease 2019 (COVID-19) among healthcare workers in Mexico City: a data-driven call to action. Clin. Infect. Dis. **73**(1), e191–e198 (2021). https://doi.org/10.1093/cid/ciaa1487
22. Zheng, Y., Zhu, Y., Ji, M., Wang, R., Liu, X., Zhang, M., Liu, J., Zhang, X., Qin, C.H., Fang, L., Ma, S.: A Learning-based model to evaluate hospitalization priority in COVID-19 pandemics. Patterns **1**(6) (2020). https://doi.org/10.1016/j.patter.2020.100092

Emotion Analysis to Identify Risk of Committing Suicide Using Statistical Learning

Luz Judith R. Esparza, Diana Barraza-Barraza, Jesús Salazar Ibarra, and Rafael G. Vargas Pasaye

Introduction

In December 2019, an outbreak of respiratory diseases caused by a new strain of the coronavirus family, designated as COVID-19, broke out. By January 20, 2020, the World Health Organization (WHO) declared a public health emergency of international concern [28] suggesting measures to avoid contagion, such as social distancing. From this day on, different countries around the world adopted social distancing, border closures, and cancelation of non-essential activities (face-to-face classes, massive events, touristic destinations, etc.).

By declaring COVID-19 a pandemic [27], the actions implemented to prevent, diagnose, and treat the infection triggered various mental health problems. Fardin [11] characterized these problems as common to infectious outbreaks, establishing that, in the face of a complete quarantine, individuals were expected to develop anxiety or depression due to losing their loved ones, food shortage, excess of negative and/or fake rumors and news, or fear of suffering from the disease.

In China, the epicenter of the epidemic, strategies have been suggested to treat the psychological impact of the COVID-19 epidemic, including the use of social networks to confront the spread of false information and improve the way psychological interventions are carried out [16]. Amid this worldwide emergency, mental health specialists face a new challenge: the work of telemedicine [17].

L. J. R. Esparza (✉)
Depto. de Matemáticas y Física Cátedra CONACyT-Universidad Autónoma de Aguascalientes, México, Mexico
e-mail: ljrodriguez@conacyt.mx

D. Barraza-Barraza
Facultad de Ciencias Exactas, Universidad Juárez del Estado de Durango, México, Mexico

J. Salazar Ibarra · R. G. Vargas Pasaye
Estudiante de Doctorado, Universidad Iberoamericana León, México, Mexico

I. Antoniano-Villalobos et al. (eds.), *Interdisciplinary Statistics in Mexico*, Springer Proceedings in Mathematics & Statistics 397,
https://doi.org/10.1007/978-3-031-12778-6_5

The care of mental health in the middle of a global emergency should not be left aside, considering the increase in suicide cases in recent years throughout the world, where most vulnerable groups are familiar with problems related to depression, anxiety, neurological damage, consumption of addictive substances, and family economic problems; conditions that a pandemic only aggravates [11].

In this context, it is important to highlight that depression is not only the presence of negative emotions, but also the absence of positive ones. The emotional reactions that have been studied the most, related to human health, have been anxiety, anger, and sadness [1]. Considering the relationships between emotions, it comes into play the usefulness of positive ones to prevent or reduce the intensity and duration of diseases, as well as to achieve high levels of individual well-being [23].

A widely used theoretical model on emotions was proposed by Plutchik [30], which presents eight basic emotions (joy, sadness, anticipation, surprise, anger, fear, disgust, and trust) with different levels of intensity that generate secondary emotions, for example, depending on the intensity of the basic emotion of sadness, the secondary emotions of grief or pensiveness could appear. This model has been the basis for previous studies on statistical learning models [32].

Therefore, considering the seriousness of this problem in the area of Public Health, the high suicide rates, as well as the consequences that suicide entails in the surviving families, this study presents a new methodology that estimates, from the written text of the individuals under analysis, the risk of committing suicide. The calculation and prediction of the index are carried out through statistical learning methodologies based on mental health experts' opinions.

This article is organized as follows: in Sect. 1, there is a relevant background on statistical learning applied to the study of suicide risk. Section 2 presents the proposed methodology to estimate the suicide risk index of Twitter users using statistical learning. The results of this research are shown in Sect. 3. Finally, the paper closes with conclusions in Sect. 4.

1 Background

Currently, there are two trends for prediction of suicidal intentions: medical and social. The latter provides predictions based on the users' behavior, with data obtained either from their interactions on social networks or from their smart devices. The constant use of social networks, as a tool that allows its users to express their feelings and thoughts, has generated various studies to consider them as a source of information about the mood of those who use them ([7, 22, 37]).

The work in [13] was among the first studies to incorporate computational techniques for the suicide risk detection and prediction finding that the level of precision in the predictions was higher when computers were used since patients preferred to answer interviews on a computer than to speak with a person, feeling more comfortable describing suicidal situations or thoughts to an entity that would not judge them. The research in [14] found, in studies dating since 2013, that Twitter is the most used social network to detect depression and other mental health issues. Christensen, Batterham, and Dea [6] analyzed the efforts in what they call e-health, or online mental health, in three areas: search for users with depression; effectiveness of interventions through social networks; and first interventions when users have been identified with risk of depression and other mental health problems.

The authors in [8] obtained public writings on Twitter of users who declared their depressive diagnosis, to compare their behavior with that of non-depressive users. Benton, Mitchell, and Hovy [3] estimated the risk of suicide in Twitter users, basing their comparison on users that have already been diagnosed with depression, while in [35], authors chose to analyze tweets through naive Bayes to detect depression.

For its part, the study in [25] used classification techniques for the analysis of texts shared over time by users. Jamil, Inkpen, Buddhitha, and White [19] developed a methodology to detect depression in Twitter users, at a tweet level. Due to the disproportionality between tweets with depressive content and those with non-depressive content, they opted for an analysis at a user level, considering the percentage of depressive tweets, improving the classification of depressed users. Also, working with Twitter, Orabi, Buddhitha, and Inkpen [29] developed classification models with high levels of accuracy of detection of whether or not a user showed signs of depression.

In [31], authors focused on the analysis of photos uploaded to Instagram to detect signs of depression in users. Using statistical methods, the authors found that it is possible to identify depressed users through the photographs they share, being able to detect signs of depression even before users are diagnosed by a mental health specialist.

More recently, Ryu, Lee, Lee and Park [33] analyzed a machine learning model to predict suicidal ideation, reaching an accuracy of 78.3%. Oh, Yun, Hwang, and Chae [26] experimented the possibility of predicting suicide attempts with a classification algorithm, reaching 77.9% detection in large samples.

Both Facebook and Twitter have implemented routines considering words and/or phrases with suicidal tone. In 2017, Facebook began using machine learning to expand the ability to provide timely help, identifying posts from people who might be at risk, looking for phrases in posts and concerned comments from friends and family [5].

The Tree Hole Rescue tool [38], developed in China to prevent suicide, tracks certain words on Weibo (the social network equivalent to Twitter) and estimates a risk of people committing suicide. From April 2018 to November 2019, the program saved 700 people. It is not an open software or with published scientific articles, so its replication is impossible to carry out.

Even though technology cannot replace mental health specialists, it can help to connect people in need with the adequate help. For now, efforts in social networks have been aimed at detecting signs of depression, without going into the study of the risk of suicide in social network users.

As can be observed from this literature review, supervised and unsupervised methods using mining text have never been used to predict suicide; therefore, this study proposes a new methodology for this gap in the literature, aiming at providing with a methodology that could be used by mental health experts in order to identify people requiring focused attention to avoid suicide attempts.

2 Materials and Methods

Statistical learning refers to a broad set of tools for understanding data. These tools can be classified as supervised or unsupervised. In general, supervised statistical learning involves building a statistical model to estimate or predict an output, based on one or more inputs. In unsupervised statistical learning, there are inputs but no outputs that "monitor" the correctness of the answer; however, relationships and structure from such data can be learned [18]. In the 1980s, Breiman, Friedman, Stone, and Olshen [4] introduced classification and regression trees, while Hastie and Tibshirani [15] coined the term Generalized Additive Models (GAMs) for a class of nonlinear extensions for Generalized Linear Models (GLMs), providing a practical software implementation. Since then, inspired by the advent of machine learning, statistical learning has evolved into a new subfield in statistics, focused on supervised and unsupervised modeling and prediction.

In this section, statistical regression and classification learning algorithms are proposed to calculate a Suicide Risk Index (SRI) through emotion mining. This study's purpose is to provide new tools for mental health experts that would help them to determine if a patient or a social networks user, who have not received mental health care, requires focused attention given their level of risk of committing suicide. The methodology consists of:

- Emotions Input: phrases written by the individual, in this case, tweets.
  The first step to process text is to standardize their format by: (1) removing non-relevant words like articles, prepositions, and conjunctions; (2) stemming the words, meaning, reducing them to their stem or root; and (3) removing punctuation signs, including http, https, # and @.
- Emotion mining: identification of emotions contained in the tweet, through NRC Word-Emotion Association Lexicon, which is a list of words and their association to the eight basic emotions [30] and to feeling classification (positive or negative) [24].
- Index calculation: Identified emotions are fed to statistical learning method to calculate the user's SRI. The SRI is a continuous value from 0 to 100 that can be further partitioned and mapped into four levels, Mild, Moderate, Severe or Extreme.

In [12], a study also using Twitter-sourced inputs, authors use the CES-D scale, that also transforms numeric responses into categories.

The following subsections provide further detail for each step in this methodology.

2.1 Emotion Mining

The phrase (tweet) input to the software is converted into an emotion profile identifying the proportion of each emotion in the phrase. For example, the phrase "His orphanhood is a terrible way to wound him, but a terrible ending is better than an endless terror" [36], an excerpt from Vega-Gil's suicide note, generates the emotions profile: Anger 18.18182%, Anticipation 9.09091%, Disgust 18.18182%, Fear 27.27273%, Joy 0%, Sadness 18.18182%, Surprise 0%, and Trust 9.09091%.

For simplification and easiness of reading, from this point on, in all mathematical notation, X_1 represents Sadness, X_2 represents Anger, X_3 stands for Fear, X_4 represents Disgust, X_5 represents Anticipation, X_6 stands for Surprise, X_7 stands for Confidence, and X_8 for Joy, the eight basic emotions presented in [30], and $\boldsymbol{X}$ is the set of eight emotions per profile, expressed as a vector, i.e., $\boldsymbol{X} = (X_1, X_2, \ldots, X_8)$.

2.2 Statistical Learning Methods

Consider the structure of the following general model:

$$SRI = f(X_1, X_2, \ldots, X_8) + \epsilon,$$

where X_j, $j = 1, 2, \ldots, 8$, is the percentage of each emotion, according to the previous section, and SRI is the response variable. The function f is a fixed but unknown function, that changes according to the learning models used in this study, and ϵ represents the random error term.

The training set for modeling and estimation of f is denoted by

$$\{(\boldsymbol{x}_1, SRI_1), (\boldsymbol{x}_2, SRI_2), \ldots, (\boldsymbol{x}_n, SRI_n)\}$$

with n, as the number of expressions that make up the training set, and $\boldsymbol{x}_i$, $i = 1, 2, \ldots, n$, is the emotion profile as defined in the Emotion Mining section, i.e., each $\boldsymbol{x}_i$ is a 8-dimensional vector.

In order to have SRI_i $(i = 1, \ldots, n)$ from 0 to 100, we have to standardized it as follows $\frac{SRI_i - \min}{\max - \min} \times 100$, where min and max are the minimum and maximum values, respectively, of all SRI_i, for $i = 1, \ldots, n$.

Subject to the existence of a response variable, statistical learning is classified into three types: Supervised (where for each $\boldsymbol{x}_i$, $i = 1, 2, \ldots, n$, there is an associated response, SRI_i), Unsupervised (for each $\boldsymbol{x}_i$, $i = 1, 2, \ldots, n$, there is no associated answer, or SRI_i), and semi-supervised which combines labeled and unlabeled data (see [39]), i.e., the practitioner can incorporate the m observations for which response measurements are available as well as the $n - m$ observations for which they are not.

Among the supervised type, there are two methods: the parametric, which assumes a functional form for f, and the non-parametric, which does not assume a functional form for f. Depending on the type of response, Regression (where SRI is presented in numerical terms) or Classification (for the case where SRI is expressed as one of the four defined levels) can be performed. In this study, supervised (both parametric and non-parametric) and unsupervised statistical learning methods are revised in order to compare their estimation/classification of SRI.

2.2.1 Supervised Models: Parametric and Non-parametric

The Generalized Additive Model (GAM) [15] serves as a parametric, supervised statistical learning method. GAMs allow nonlinear relationships between the SRI response and the set of emotions, through the following model structure:

$$SRI = \beta_0 + g_1(X_1) + g_2(X_2) + \cdots + g_8(X_8) + \epsilon,$$

where $g_j(X_j)$, $j = 1, 2, \ldots, 8$, is a nonlinear, smooth function, estimated separately for each emotion, their contributions are added to give a numerical SRI. According to Hastie & Tibshirani [15], the g_j may be a function estimated parametrically, non-parametrically, or semi-parametrically. This study uses the former one.

From the non-parametric models, the spline model [18] was selected, given the advantages provided by adjusting curves by regions, that is, a curve is adjusted depending on the values taken by the emotion profile. These models are given by:

$$SRI = \beta_0 + \beta_1 b_1(X_1) + \beta_2 b_2(X_2) + \cdots + \beta_8 b_8(X_8) + \epsilon,$$

where $b_j(X_j)$, $j = 1, 2, \ldots, 8$ is a fixed, polynomic function of X_j.

2.2.2 Unsupervised Model: Lineal Discriminant Analysis

Predicting a qualitative response is known as observation classification, since it involves assigning the observation to a category or class. For this study, it is assumed that the emotion profile, $\boldsymbol{X} = (X_1, \ldots, X_8)$, has a multivariate Gaussian distribution (or multivariate normal), with a vector of means $\boldsymbol{\mu}_k$ (with $k = 4$, the number of SRI levels, that is: Mild, Moderate, Severe, and Extreme) and a common covariance matrix in all classes $\boldsymbol{\Sigma}$. The observations can come from one of four Gaussian distributions (represented as a mixture model). The density of $\boldsymbol{X}$ is given by

$$f_{\boldsymbol{X}}(\boldsymbol{x}) = \frac{1}{(2\pi)^4 |\boldsymbol{\Sigma}|^{(1/2)}} \exp\left(-\frac{1}{2}(\boldsymbol{x} - \boldsymbol{\mu}_k)^\top \boldsymbol{\Sigma}^{-1}(\boldsymbol{x} - \boldsymbol{\mu}_k)\right).$$

Using this information, Linear Discriminant Analysis (LDA) [4] determines to which of the four levels of SRI belongs an emotion profile, using the following equation:

$$\delta_k(\boldsymbol{x}) = \boldsymbol{x}^\top \boldsymbol{\Sigma}^{-1} \boldsymbol{\mu}_k - \frac{1}{2}\boldsymbol{\mu}_k^\top \boldsymbol{\Sigma}^{-1} \boldsymbol{\mu}_k + \log \pi_k,$$

where π_k is the proportion of training observations that belong to k-th SRI level. In other words, $\pi_k = n_k/n$, where n_k is the number of elements at SRI level k. These numbers are determined using information from the training set.

2.3 Training and Test Datasets

Predicting suicide through text mining is not an easy task, since there are different mechanisms such as grammatical errors, slang, and other lexical characteristics that influence the analysis of this mental disorder. Therefore, since a group of mental health experts has already contributed to analyze emotion profiles and linked them to a suicide index (see [32]), this study uses their dataset for training and testing the proposed methodology.

For the study conducted in [32], mental health experts were provided with a series of emotion profiles and asked to assign a suicide index, according to their experience. From this exercise, a dataset with 177 observations was generated, from which, 100 were used as a training dataset, and the other 77 as a test set, in the analysis presented in the following section.

3 Results

The information provided by the mental health experts in [32] was used as the training data (100 data) set for the statistical learning algorithms used in this study. Given that the experts provide numerical information and no risk level (mild, moderate, severe or extreme), a hierarchical cluster analysisv with the average group linkage method was performed, in order to define limits for the risk groups.

It should be noted that all the implementations of the models were performed in the statistical programming language R. The results are presented in Fig. 1, with the proposed intervals for SRI levels as: Mild 0–45, Moderate 46–70, Severe 71–90, and Extreme 91–100.

3.1 Supervised Models

Regression models were fitted to the training set. Table 1 presents the analysis of variance (ANOVA) for the parametric model (linear regression) showing significance the variables Sadness, Anger, Fear, Disgust, and Anticipation (marked with ***), for the response variable SRI.

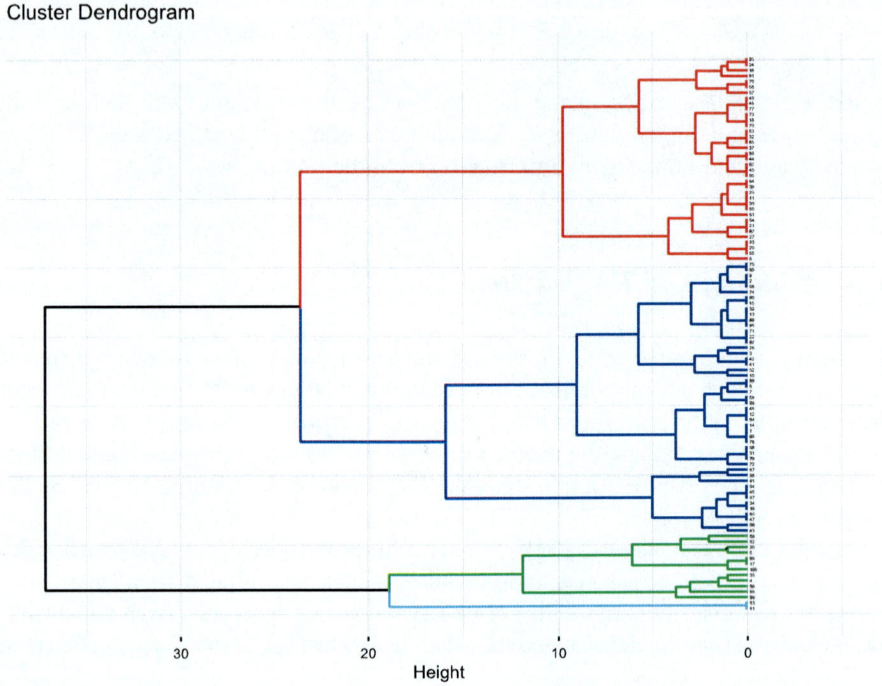

Fig. 1 Dendrogram of training data using a Hierarchical Clustering Algorithm

Table 1 ANOVA for the parametric model to estimate the SRI

	Df	Sum Sq	Mean Sq	F value	Pr(>F)	
Sadness	1	22736.9	22736.9	956.047	>2.20E-16	***
Anger	1	9356.8	9356.8	393.438	> 2.20E-16	***
Poly (Fear, 2)	2	1069.3	534.6	22.480	1.086E-08	***
Disgust	1	643.2	643.2	27.047	1.176E-06	***
Anticipation	1	305.4	305.4	12.843	0.0005416	***
Residuals	93	2211.7	23.8			

Table 2 presents the ANOVA table for natural splines (ns) model fitted to SRI training dataset, using different degrees of freedom (shown in parenthesis in the first column), being significant, for this model, the variables Sadness, Anger, Fear, Disgust, and Anticipation (marked with *** and **).

Figure 2 presents QQ plots for the residuals of both, parametric and non-parametric, models. Shapiro-Wilk normality test was used for the residuals, obtaining, for the parametric model, a statistic of $W = 0.98297$ and a p-value $= 0.2248$; whereas for the non-parametric model, $W = 0.98125$ and p-value $= 0.1662$. The

Table 2 ANOVA for the non-parametric model to estimate the SRI

	Df	Sum Sq	Mean Sq	F value	Pr(>F)	
Ns (Sadness, 3)	3	23845.2	7948.4	360.8142	>2.20E-16	***
Ns (Anger, 3)	3	8351.7	2783.9	126.3740	>2.20E-16	***
Ns (Fear, 2)	2	1180.4	590.2	26.7914	8.591E-10	***
Ns (Disgust, 2)	2	775.2	387.6	17.5955	3.827E-07	***
Ns (Anticipation, 2)	2	254.4	127.2	5.7731	0.004423	**
Residuals	87	1916.5	22.0			

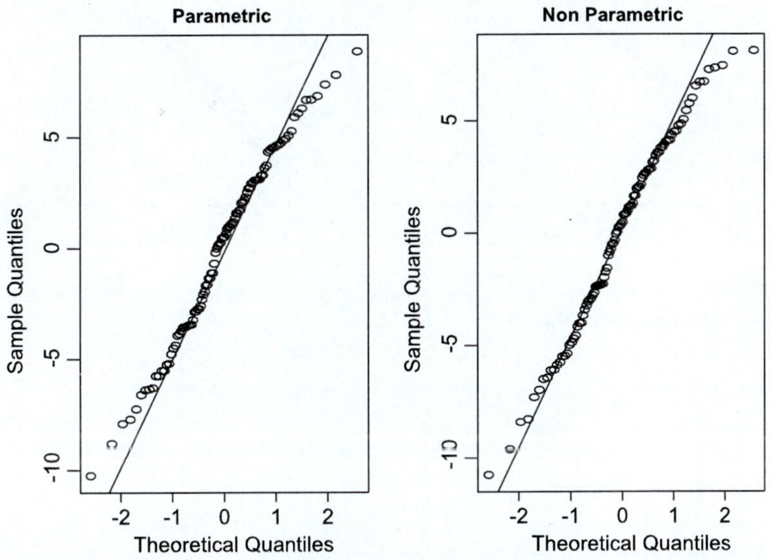

Fig. 2 QQ plot of the residuals using the parametric and the non-parametric models

QQ plots and Shapiro–Wilk test showed no evidence to reject normality on both sets of residuals.

Figure 3 presents the real SRI for the training dataset and its corresponding estimation, including the confidence intervals, performed with parametric and non-parametric models. The reader should notice that the SRI values are sorted in this figure and that both methods show good performance in their corresponding SRI estimation.

In order to perform a comparison of the parametric and non-parametric models, using a published dataset [32], 77 phrases of Twitter users were considered as the test set for SRI estimation. The results are presented in Fig. 4; from the figure, it can be observed that both models estimate similar SRI levels (numeric and categorical).

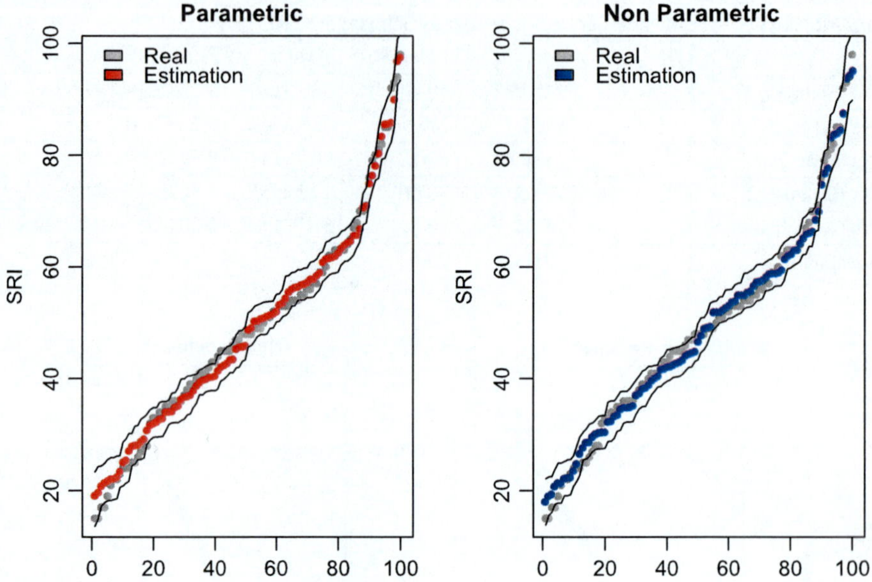

Fig. 3 Sort SRI training data and their estimation (including confidence intervals) using the parametric and the non-parametric models

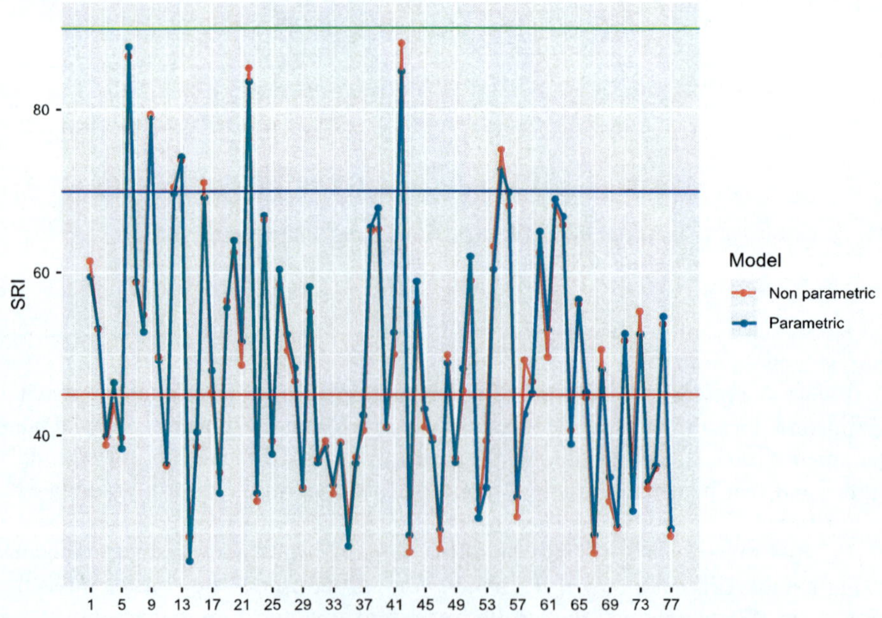

Fig. 4 Prediction of the SRI using test data. Comparison of the parametric and non-parametric models

Table 3 Confusion matrix for 20% of the training dataset using a Naive Bayes classifier and LDA

	Prediction							
	Mild		Moderate		Severe		Extreme	
Real	Bayes	LDA	Bayes	LDA	Bayes	LDA	Bayes	LDA
Mild	7	8	0	1	0	0	0	0
Moderate	1	0	10	9	0	0	0	0
Severe	0	0	0	0	0	0	0	1
Extreme	0	0	0	0	0	0	2	1

3.2 *Unsupervised Model*

Using the training set, the library *mlr3* from R (a modern object-oriented machine learning framework), and cross validation using $k = 10$ folds, the accuracy (shown in parenthesis) of the following classifiers was compared: debug (0.30), featureless (0.34), rpart (0.73), KKNN (0.80), LDA (0.83), Naive Bayes (0.90), xgboost (0.82), and SVM (0.82). For more detail on these classifiers, see [2, 20]. Given the accuracy shown by LDA and Naive Bayes, this study deepens in their performance for SRI classification. The training dataset was split into two sets: 80% served as a new training set, and the classifiers Naive Bayes and the LDA (this is an example of supervised classification, see [18]) were tested; the other 20% served as a test set for construction of the confusion matrix presented in Table 3. Obtaining an accuracy for the Naive Bayes classifier of 95% and for the LDA is 90%. However, due to its simplicity and the fact that, as indicated in [10] and [21], the LDA frequently achieves good performances even though the assumptions of common covariance matrix among groups and normality is often violated, this study opted for this classifier instead of Naive Bayes.

The LDA model estimates the probabilities of belonging to each SRI level: Mild 0.45, Moderate 0.43, Severe 0.08, and Extreme 0.04. From the training data, the SRI levels with the highest probabilities are mild or moderate.

The LDA describes the groups through a vector of means (rows in Table 4). Remembering that these are percentages of emotions identified in each profile, it is possible to observe that the *mild risk level* is represented by low percentages of negative emotions such as sadness, anger, fear or disgust, and high levels of trust and joy. On the other hand, the *extreme risk level* is identified by high percentages of negative emotions (sadness, anger, fear, and disgust), while positive emotions present very low or zero levels (joy, surprise, and anticipation).

Table 4 Mean vectors for the SRI levels using the LDA

	Sadness	Anger	Fear	Disgust	Anticipation	Surprise	Trust	Joy
Mild	6.234	2.578	7.009	4.399	16.965	7.704	31.976	23.132
Moderate	14.636	10.277	15.604	8.544	14.978	6.753	18.487	10.719
Severe	30.565	14.018	16.585	17.345	6.637	3.750	6.013	5.085
Extreme	32.618	20.737	24.947	20.382	0.000	0.000	1.315	0.000

3.3 Model Comparison

This section compares the three previous methods (parametric, non-parametric and unsupervised) using the test the test dataset. Additionally, the Analytic Hierarchy Process (AHP), presented in [32] for analysis of emotions, was used as a mean to compare the methodology to already published studies.

Figure 5 presents how many users of the test data were classified in each category for the three methods. The unsupervised and parametric supervised models coincide on approximately 92% of the cases. The supervised models present a higher percentage of estimation coincidences, with approximately 94% of coincidence on the input phrases.

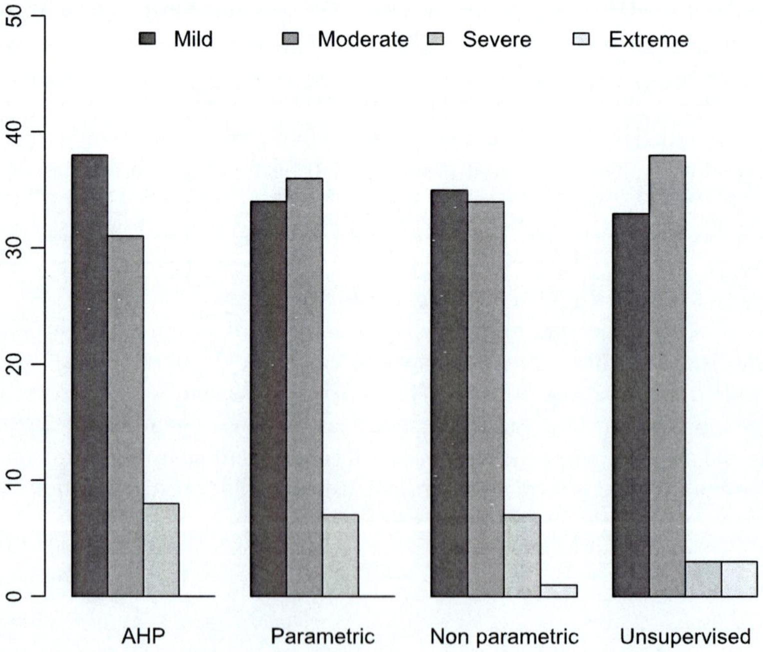

Fig. 5 Classification of the test data obtained from the four models at each SRI level

3.4 Testing the Models with a New Test Set: COVID-19

Considering the pandemic status in the world due to COVID-19 that imposes a social distancing strategy to control the coronavirus, the purpose of this study is to propose the methodology as a way to monitor mental health from a distance, exemplifying it with Twitter users; therefore, this methodology is applied to analyze the emotions of various leaders from Mexico, expressed in their tweets between March 15 to April 12, 2020; analyzing 2,377 words for leaders of the Mexican Republic.

The opinion of the political leaders is of the utmost importance to society, since from their expression, the general population can base their own opinion and, in an scenario like COVID-19, even bias their feelings and attitude toward the pandemic ([9, 34]).

Figure 6 shows the results obtained by applying the supervised models (parametric and non-parametric) on tweets from Mexican leaders, where can be seen that most of them show an SRI below 60. The only leader with an SRI above 70 is from the State of Hidalgo.

When comparing the results of the supervised models with the unsupervised one, as shown in Table 5, it can be seen that Mexican leaders show SRI levels that do not trigger any alarm, since the tweets show that these individuals maintained a positive active attitude trying to convey the risks of the pandemic; however, for the leader of

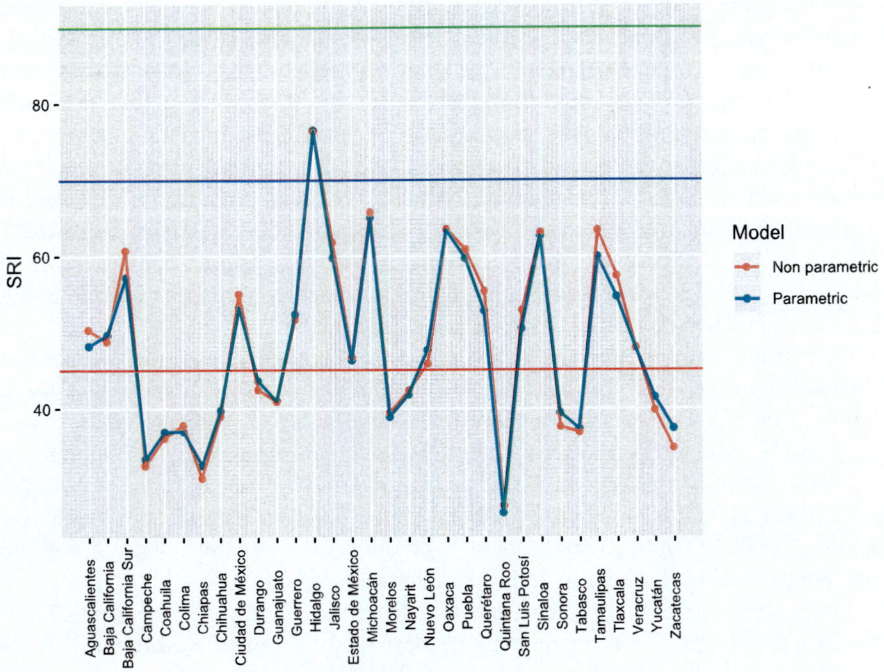

Fig. 6 SRI predictions using supervised models for Mexican leaders

Table 5 Mexican leaders classified under each SRI level according to supervised and unsupervised models

	SRI			
Model	Mild	Moderate	Severe	Extreme
Parametric	14	17	1	0
Non-Parametric	14	17	1	0
Unsupervised	14	17	0	1

Hidalgo, the supervised models predict severe levels, while the unsupervised model predicts an extreme level for his SRI. The emotion analysis perform to this leader's tweets showed that his emotions leaned toward sadness and fear.

4 Conclusions

This article proposed a new methodology to identify suicidal intentions of Twitter users, based on what they express on this platform, through statistical learning methods, using three models: parametric, non-parametric, and LDA. The training dataset was generated at a national meeting of mental health experts, and the models were tested with tweets retrieved from generic users. Additionally, the analysis includes tweets from leaders from Mexico issued after the declaration of the COVID-19 pandemic.

The classification performed to the test set by the three models gave the same results in above 90% of the phrases analyzed. The parametric model did not predict or identify any phrase (user) with extreme SRI level; non-parametric model predicted one user, while the LDA (unsupervised model) predicted three users. However, it is important to note that both the parametric and the non-parametric classified 7 users with severe SRI levels, three of them are those that appear as extreme when using LDA. Thus, we conclude that the supervised methods are more relaxed in the classification when compared to the unsupervised method.

Comparing the performance of the proposed methods against the estimations obtained from the AHP, results show that the number of users in each category is similar; however, carrying out a more exhaustive analysis, with the AHP, it was not possible to identify any user with extreme suicidal intentions. It is important to highlight that it is considered a better result to classify a user in higher risks than those in reality, that is, it is better to obtain a false positive that indicates that a user requires attention, and provide it, than to obtain a false negative, and not provide mental health care, resulting in an irreparable loss.

Given the nature of the analyzed data in this article, it is important to mention that misclassification problems are inherent to these methods, and therefore, the authors would like to emphasize that the tools designed here are not intended as a replacement of mental health specialists, but rather, are proposed as a mean to help them identify individuals at high risk of attempting suicide. It remains as future work to make this methodology available to mental health experts in the form of an app.

Acknowledgements Esparza is grateful for the support from CONACyT through a Cátedra at the Autonomous University of Aguascalientes. Also, the authors would like to express their most sincere acknowledgment to the referees who have contributed to the improvement of this article.

References

1. Abascal, E.G.F., Palmero, F.: Emociones y salud. Ariel (1999). http://reme.uji.es/articulos/ajimxm7392902100/texto.html
2. Becker, M., Binder, M., Bischl, B., Lang, M., Pfisterer, F., Reich, N.G., Richter, J., Schratz, P., Sonabend, R., Pulatov, D.: Mlr3 book (2021). https://mlr3book.mlr-org.com
3. Benton, A., Mitchell, M., and Hovy, D.: Multi-task Learning for Mental Health using Social Media Text (2016). arXiv preprint ArXiv:1712.03538
4. Breiman, L., Friedman, J., Stone, C.J., Olshen, R.A.: Classification and Regression Trees. CRC Press (1984)
5. Card, C.: How Facebook AI Helps Suicide Prevention. Facebook Newsroom (2018)
6. Christensen, H., Batterham, P.J., Dea, B.O.: E-health interventions for suicide prevention. Int. J. Environ. Res. Publ. Health. **11**(8), 8193–8212 (2014). https://doi.org/10.3390/ijerph110808193
7. Dang, N.C., Moreno-García, M.N., De la Prieta, F.: Sentiment analysis based on deep learning: A comparative study. Electronics **9**(3), 483 (2020)
8. De Choudhury, M., Gamon, M., Counts, S., Horvitz, E.: Predicting depression via social media. In Proceedings of the Seventh International AAAI Conference on Weblogs and Social Media Predicting **2**, 128–137 (2013)
9. Doerr, A.J.: Locked (down) and loaded (language): Effect of policy and speech on COVID-19 outcomes. J. Leadersh. Organ. Stud. **28**(3), 340–348 (2021). https://doi.org/10.1177/15480518211012404
10. Duda, R.O., Hart, P.E.: Pattern Classification. John Wiley & Sons (2006)
11. Fardin, M.A.: COVID-19 and Anxiety: A Review of Psychological Impacts of Infectious Disease Outbreaks. *Archives of Clinical Infectious Diseases, (In Press)*. (2020)
12. Gómez, G.A.: Evaluación del riesgo de suicidio: enfoque actualizado. Revista Médica Clínica Las Condes. **23**(5), 607–615 (2012)
13. Greist, J.H., Gustafson, D.H., Satuss, F., Rowse, G.L., Laughren, T.P., Chiles, J.A.: A computer interview for suicide-risk prediction. Am. J. Psychiatry **130**(12), 1327–1332 (1973)
14. Guntuku, S.C., Yaden, D.B., Kern, M.L., Ungar, L.H., Eichstaedt, J.C.: Detecting depression and mental illness on social media: an integrative review. Curr. Opin. Behav. Sci. **18**, 43–49 (2017)
15. Hastie, T., Tibshirani, R.: Generalized additive models: some applications. J. Am. Stat. Assoc. **82**(398), 371–386 (1987)
16. Ho, C.S., Chee, C.Y., Ho, R.C.: Mental health strategies to combat the psychological impact of COVID-19 beyond paranoia and panic. Ann. Acad. Med. Singapore **49**(1), 1–3 (2020)
17. Hollander, J.E., Carr, B.G.: Virtually perfect? Telemedicine for covid-19. N. Engl. J, Medi (2020)
18. James, G., Witten, D., Hastie, T., Tibshirani, R.: An Introduction to Statistical Learning: with Applications in R, 7th edn. Springer, New York (2017)
19. Jamil, Z., Inkpen, D., Buddhitha, P., White, K.: Monitoring Tweets for Depression to Detect At-risk Users. Université d'Ottawa, University of Ottawa (2017)

20. Lang, M., Binder, M., Richter, J., Schratz, P., Pfisterer, F., Coors, S., Au, Q., Casalicchio G., Kotthoff, L., Bischl, B.: Mlr3: A modern object-oriented machine learning framework in R. J. Open Source Softw. (2019). https://joss.theoj.org/papers/10.21105/joss.01903
21. Li, T., Zhu, S., Ogihara, M.: Using discriminant analysis for multi-class classification: an experimental investigation. Knowl. Inf. Syst. **10**(4), 453–472 (2006)
22. Liu, B., Tang, S., Sun, X., Chen, Q., Cao, J., Luo, J., Zhao, S.: Context-aware social media user sentiment analysis. Tsinghua Sci. Technol. **25**(4), 528–541 (2020)
23. Lyubomirsky, S., King, L., Diener, E.: The benefits of frequent positive affect: Does happiness lead to success? Psychol. Bull. **131**(6), 803 (2005)
24. Mohammad, S.M., Turney, P.D.: Crowdsourcing a word-emotion association lexicon. Comput. Intell. **29**(3), 436–465 (2013)
25. Nadeem, M.: Identifying depression on Twitter (2016). arXiv Preprint ArXiv:1607.07384
26. Oh, J., Yun, K., Hwang, J.-H., Chae, J.-H.: Classification of suicide attempts through a machine learning algorithm based on multiple systemic psychiatric scales. Front. Psychiatry **8**, 192 (2017)
27. OMS.: WHO Director-General's opening remarks at the media briefing on COVID-19–11 March 2020 (2020a). Retrieved from https://www.who.int/dg/speeches/detail/who-director-general-s-opening-remarks-at-the-media-briefing-on-covid-19---11-march-2020
28. OMS.: WHO Director-General's statement on IHR Emergency Committee on Novel Coronavirus (2019-nCoV). Geneva, World Health Organization (2020b). Retrieved from https://www.who.int/dg/speeches/detail/who-director-general-s-statement-on-ihr-emergency-committee-on-novel-coronavirus-(2019-ncov)
29. Orabi, A.H., Buddhitha, P., Orabi, M.H., Inkpen, D.: Deep learning for depression detection of twitter users. In: Proceedings of the Fifth Workshop on Computational Linguistics and Clinical Psychology: From Keyboard to Clinic, pp. 88–97 (2018)
30. Plutchik, R.: The nature of emotions: Human emotions have deep evolutionary roots, a fact that may explain their complexity and provide tools for clinical practice. Am. Sci. **89**(4), 344–350 (2001)
31. Reece, A.G., Danforth, C.M.: Instagram photos reveal predictive markers of depression. IEPJ Data Sci. **6**(1), 15 (2017)
32. Rodríguez-Esparza, L.J., Barraza-Barraza, D., Salazar-Ibarra, J., Vargas-Pasaye, R.G.: Index of suicide risk in Mexico using twitter. J. Soc. Res. **5**(15), 1–13 (2019)
33. Ryu, S., Lee, H., Lee, D.-K., Park, K.: Use of a machine learning algorithm to predict individuals with suicide ideation in the general population. Psychiatry Inv. **15**(11), 1030 (2018)
34. Sergent, K., Stajkovic, A.D.: Women's leadership is associated with fewer deaths during the COVID-19 crisis: Quantitative and qualitative analyses of United States governors. J. Appl. Psychology **105**(8), 771–783 (2020). https://doi.org/10.1037/apl0000577
35. Shen, G., Jia, J., Nie, L., Feng, F., Zhang, C., Hu, T.,. . . , Zhu, W.: Depression detection via harvesting social media?: A multimodal dictionary learning solution. In: Proceedings of the Twenty-Sixth International Joint Conference on Artificial Intelligence, pp. 3838–3844. (2017)
36. Vega-Gil, A.: Nota Suicida (2019). Retrieved April 5, 2019, From https://twitter.com/ArmandoVegaGil/status/1112666222951391233
37. Wang, L., Niu, J., Yu, S.: Sentidiff: Combining textual information and sentiment diffusion patterns for twitter sentiment analysis. IEEE Trans. Knowl. Data Eng. **32**(10), 2026–2039 (2019)
38. Want, Y.: The Chinese suicides prevented by AI from afar. BBC News (2019). Retrieved from https://www.bbc.com/news/technology-50314819
39. Zhu, X., Goldberg, A.B.: Introduction to semi-supervised learning. Synth. Lect. Artif. Intell. Mach. Learn. **3**(1), 1–130 (2009). https://doi.org/10.2200/S00196ED1V01Y200906AIM006

Characterizing Groups Using Latent Class Mixed Models: Antiretroviral Treatment Adherence Analysis

Ruth Fuentes-García, Lizbeth Naranjo, and Yanink Caro-Vega

1 Introduction

Current Antiretroviral Treatment (ART) regimes allow people with HIV (PWH) to achieve and maintain viral suppression up to three to five years since the first ART initiation [1–3]. Adherence to the treatment, that is, opportune and complete intake of the prescribed dosage, is of crucial importance for the patient's evolution. Incomplete adherence has been identified as one of the potential factors that could drive heightened inflammation, immune activation, and coagulopathy in adult people with HIV receiving ART [4–6]. Hence, understanding the adherence dynamics across time for PWH is relevant for the adequate surveillance of their condition.

The linear mixed model is frequently used to describe change over time of a dependent Gaussian variable, often a set of covariates is considered to determine their possible effect over such change. A particular challenge in our problem at hand is the presence of heterogeneity in the population, where subpopulations with different growth curve shapes could fit better. Namely, we are interested in identifying different groups of patterns for the dependent variable. This in turn would define subpopulations with similar patterns among the observations. Latent Class Mixed Models (LCMM), provide a useful proposal to tackle the problem. These models combine the characteristics of a linear mixed model with a latent variable, which divides the population into subsets of homogeneous observations. To estimate the parameters in these models, we used the library lcmm available in R [7]. Muthén [8, 9] proposed a finite mixture model with random effects known in the literature as Growth Mixture Models, as a tool to represent unobserved heterogeneity between

R. Fuentes-García · L. Naranjo (✉) · Y. Caro-Vega
Facultad de Ciencias, Universidad Nacional Autónoma de México, Mexico City, Mexico
e-mail: lizbethna@ciencias.unam.mx

Y. Caro-Vega
Departamento de Infectología, Instituto Nacional de Ciencias Médicas y Nutrición Salvador Zubirán, Mexico City, Mexico

I. Antoniano-Villalobos et al. (eds.), *Interdisciplinary Statistics in Mexico*,
Springer Proceedings in Mathematics & Statistics 397,
https://doi.org/10.1007/978-3-031-12778-6_6

observations, hence capturing latent trajectory classes with different growth curve shapes. His approach is widely used and is implemented in the licensed software Mplus, it uses a continuous latent variable, defined as a function of the covariables of interest with class-specific parameters and individual random effects. Given the class membership, this model leads to different multivariate Normal distributions than those defined by the LCMM, however, in general, it is similar in purpose and construction.

The HIV viral load and CD4+T lymphocytes counts are the most used biomarkers to monitor HIV-positive patients. The HIV infection progression is characterized by a progressive decrease of CD4+T counts and an increase of CD8+T counts, and therefore, a simultaneous decrease of CD4/CD8 ratio. The CD4/CD8 ratio has been used as a biomarker of disease stage in PWH, as a marker of immune activation and immune senescence. Persistently, low CD4/CD8 ratio has been associated with residual inflammation and identified as a good predictor of increased risk of death [10]. The CD4/CD8 ratio might represent a good predictor of AIDS and non-AIDS events, it is a more widely available measurement than other inflammatory biomarkers and is part of routine clinical care when monitoring CD4+T in many clinical settings. However, low CD4/CD8 ratios could be observed for different combinations of values of the numerator and denominator of the ratio. It would, therefore, be useful to explore the joint behavior of CD4+T and CD4/CD8 ratio. Antiretroviral adherence in people with HIV and virological suppression is associated with a decreased CD4-CD8 ratio [11].

The outline of this paper is organized as follows. In Sect. 2, we describe the population under study and the adherence to treatment. Section 3 describes the general setting for the latent class mixed models. In Sect. 4, we show the results and the classes identified through the LCMM; and finally a discussion and conclusions are given in Sect. 5.

2 Materials

In this section, we describe the population, the data sources, and data collection used for this study, and the ART adherence definition.

2.1 Population

A subset of PWH receiving medical care at the HIV/AIDS Clinic at the Instituto Nacional de Ciencias Médicas y Nutrición Salvador Zubirán (INCMNSZ) is considered, which is a national referral center located in Mexico City. Patients included in this analysis were those diagnosed with HIV-infection, that started their first antiretroviral treatment (ART) regimen between January 2005 and September 2017, were followed up for at least six months and achieved viral suppression. The viral suppression is defined as having less than 200 copies of HIV per milliliter of blood (copies/μL). These patients were followed only if they showed uninterrupted periods of HIV suppression and if adherence information was available. Therefore, time in follow-up started the date of the first HIV RNA measurement <200 copies/μL after having started ART. Follow-up ended the date of a confirmed HIV RNA measurement $\geq$400 copies/μL, a period longer than six months without ART provision, last medical visit, death, or cohort administrative censoring (September 28th, 2017). For this analysis, the subset of the population we considered corresponds to 988 patients in the first five years of their follow-up period.

2.2 Data Collection

Data from the HIV/AIDS clinic and pharmacy databases was retrieved. Additionally, prospective information on all PWH receiving care in clinic was collected. The database contains patients' demographic, clinical, and laboratory characteristics. Data are collected at each visit and entered immediately into the database. As a clinical HIV-care center, receiving funds for ART from Federal Government's Ministry of Health, the clinic provides free-of-charge ART prescribed by the physicians in the clinic [12]. Patients refill their medications monthly, or bimonthly, and the pharmacy records electronically the ART provision. The pharmacy registers the refill date and when 30- or 60-day provision of ART is delivered to each patient.

The CD4+T and CD8+T lymphocyte counts (cells/μL) are measured by flow cytometry, at baseline (at enrollment) and routinely two or three times a year, during follow-up visits [13]. The CD4/CD8 ratio is computed from the division of the CD4+T lymphocyte counts (cells/μL)over the CD8+T lymphocyte counts (cells/μL).

2.3 ART Adherence Definition

We measured adherence (ADH) using the medication possession ratio (MPR) [14] based on pharmacy registries. We divided the 30-day doses of ART provided for a month period by the total number of days that the patient took to refill their next prescription to calculate the percentage of days that patients have ART on their possession. All surplus ART were taken into account in case the patients had picked up a previous prescription prior to 30 days. For example, if a patient filled his/her prescription at day 25, she/he was considered to have a 100% adherence for that month, and we carried forward the extra five-day provision to the next month's estimated MPR [14]. The moving average of the last four months was taken as the measure of adherence for the statistical analysis [15]. Adherence was capped at 100%.

3 Latent Class Mixture Models

The latent class mixed model considers a heterogeneous population composed of K latent classes of subjects characterized by K mean profiles of trajectories. Each subject is assigned to one and only one latent class. The number of classes should be established a priori by choosing the best fit and comparing the performance of models with different number of subpopulations, commonly through likelihood-based criteria. The model for subject i with n_i repeated measurements is given as follows:

$$\mathbf{Y}_i = \mathbf{X}_i\boldsymbol{\beta} + \mathbf{W}_i\boldsymbol{\lambda}_k + \mathbf{Z}_i\mathbf{b}_i + \mathbf{e}_i,$$

where $\mathbf{Y}_i$ is the response vector for subject i with his/her n_i observations along time, $\mathbf{X}_i$ is the design matrix of fixed effects and $\mathbf{Z}_i$ is the design matrix for random effects; $\boldsymbol{\beta}$ is the vector of regression coefficients for the fixed effects; $\mathbf{b}_i$ is the random effects vector; and random error terms corresponds to $\mathbf{e}_i$; the latter two terms are assumed mutually independent with mean $\mathbf{0}$ with $\text{Var}(\mathbf{b}_i) = \mathbf{B}$ and $\text{Var}(\mathbf{e}_i) = \mathbf{R}_i$. Finally, class-specific fixed effects are included in the model through the term $\mathbf{W}_i\boldsymbol{\lambda}_k$, where $\mathbf{W}_i$ and $\boldsymbol{\lambda}_k$ are the class-specific fixed effects and parameter vector for class k, respectively. Note that $\boldsymbol{\beta}$ will apply to all values of the corresponding column in $\mathbf{X}_i$, while $\boldsymbol{\lambda}_k$ will only affect observations in class k. For subject i, dummy variables are defined

$$c_{ik} = \begin{cases} 1 & \text{if subject } i \text{ belongs to class } k, \\ 0 & \text{otherwise,} \end{cases}$$
$$(c_{i1}, \ldots, c_{iK}) \sim \text{Multinomial}\left(1, (\pi_{i1}, \ldots, \pi_{iK})\right),$$

where $\boldsymbol{c}_i = (c_{i1}, \ldots, c_{iK})'$. Probabilities π_{ik} are given by:

$$\pi_{ik} = \mathbb{P}(c_{ik} = 1|\mathbf{v}_i) = \frac{\exp(\mathbf{v}_i'\boldsymbol{\alpha}_k)}{\sum_{j=1}^{K}\exp(\mathbf{v}_i'\boldsymbol{\alpha}_j)},$$

$\mathbf{v}_i$ are variables used to determine the class membership for subject i and $\boldsymbol{\alpha}_j$'s are the unknown parameters vectors associated to class assignations, with constraint $\boldsymbol{\alpha}_1 = \mathbf{0}$.

The likelihood given the observed design matrices could be written as

$$\log L(\theta) = \sum_{i=1}^{n}\log\sum_{k=1}^{K}\pi_{ik} f\left(y_{ik}|c_{ik} = 1\right),$$

where $f(y_{ik}|c_{ik} = 1)$ is the density of a $\text{N}(\mathbf{X}_i\boldsymbol{\beta} + \mathbf{W}_i\boldsymbol{\lambda}_k\,,\ \mathbf{Z}_i\mathbf{B}\mathbf{Z}_i' + \mathbf{R}_i)$.

This model could be extended to describe multiple observed longitudinal markers. In this case, an underlying latent process $\Lambda_i(t)|c_{ik}$ is assumed, generated by the d longitudinal marker

$$\Lambda_i(t)|c_{ik} = \mathbf{X}_i(t)'\boldsymbol{\beta} + \mathbf{W}_i(t)'\boldsymbol{\lambda}_k + \mathbf{Z}_i(t)'\boldsymbol{b}_i + \omega_i(t),$$

where $\{\omega_i(t)\}$ is a zero mean Gaussian stochastic process. Let Y_{dij} be the measure of marker d, $d = 1, \ldots, D$, for subject i, $i = 1, \ldots, N$, at occasion j, $j = 1, \ldots, n_{di}$. The corresponding measurement of time is denoted t_{dij}. Here

$$Y_{dij} = H_d(\tilde{Y}_{dij}; \boldsymbol{\eta}_d), \tag{1}$$

where H_d is a parameterized marker-specific link function, $\boldsymbol{\eta}_d$ are the parameters for H_d; in this work, linear link functions were used, i.e., $H^{-1}(Y) = \frac{Y-\eta_1}{\eta_2}$, where η_1 and η_2 are location and scale parameters, respectively. And $\tilde{Y}_{dij}$ is the noisy latent process at time t_{dij},

$$\tilde{Y}_{dij} = \Lambda_i(t_{dij})|c_{ik} + \mathbf{U}_i(t_{dij})'\gamma_d + \delta_{di} + \epsilon_{dij},$$

where $\mathbf{U}_i$ are covariates with a marker-specific effect γ_d called contrast since $\sum_{d=1}^{D} \gamma_d = 0$, δ_{di} is a random intercept, and $\epsilon_{dij} \sim N(0, \sigma^2_{\epsilon_d})$ is the independent random measurement error. In this case, the individual contribution to the likelihood would consider the Jacobian determinant of the link function and the density of a multivariate normal variable.

Optimizing the log-likelihood for mixture models could be done using EM algorithms or Newton–Raphson type algorithms. For this analysis, we used the LCMM library available in R [7] which estimation uses maximum likelihood theory with a powerful modified Marquardt iterative algorithm, and a Newton–Raphson like algorithm. These choices were based on the previous analyses regarding the speed and convergence of the method. A set of initial values is required, and the choice could affect the convergence, hence to ensure whether the global maximum is reached, they suggest using several values. The choice of such values could be made through a grid or using random draws from one class estimated model. It is worth mentioning that several values should be tried to ensure that there are no convergence concerns.

Once the models are fitted using the lcmm library, a quick comparison between them could be done using library LCTMtools. The common criteria used to compare models are Akaike information criteria (AIC) [16], Bayesian information criteria (BIC) [17], and entropy [18] (a criterion to assess the quality of the classification), where the best model is that having the lowest AIC and BIC, and the highest entropy.

4 Results

In this section, we show the results of the LCMM. In the first stage, the outcome was the ART adherence over time, and LCMM is used to find groups of ART adherence trajectories. Then, in the second stage, within each adherence group, we considered a bivariate outcome, namely the CD4/CD8 ratio and the CD4+T counts in their

Table 1 AIC, BIC, entropy and relative entropy for LCMM with different number of latent classes for ART adherence

Number of classes	AIC	BIC	Entropy	Relative entropy
1	66931.8	66975.8		
2	66433.5	66492.2	0039.08	0.94
3	66439.5	66512.2	0629.16	0.42
4	66445.5	66533.6	1030.25	0.25

logarithmic scale, and the LCMM is used to explain the biomarkers behavior by using some characteristics of the patients as explanatory variables.

4.1 *Latent Classes from ART Adherence*

In this section, we describe the population characteristics at baseline and the ART adherence classes. The analyzed data has 988 patients (894 male and 94 female) with a median follow-up of 4.13 (interquartile range, IQR, 1.87–7.18) years; their median ART adherence was 99.17% and IQR 95.83%–100%, having an asymmetric distribution; their median age at the beginning of follow-up was 33 and IQR 28–41; 287 (29%) patients were diagnosed with AIDS-related conditions indicator previous to enrollment in care (if the patient was diagnosed with any of the opportunistic infections or other specific illnesses marked as AIDS-defining events by the CDC[1] at an earlier date or at the time of enrollment); the CD4+T counts at the beginning of follow-up were categorized as $\leq$200, 201–500 and >500 cells/mm^3 (341, 469 and 178 patients, respectively); their median CD4+T counts at the beginning of follow-up was 288.5 (IQR 158.0–443.8) and at the end of follow-up 484.0 (IQR 337.0–647.2); their median CD8+T counts at the beginning of follow-up was 847.0 (IQR 574.0–1175.2) and at the end of follow-up 727.0 (IQR 545.8–990.0); their median CD4/CD8 ratio at the beginning of follow-up was 0.329 (IQR 0.189–0.534) and at the end of follow-up 0.641 (IQR 0.431–0.914).

LCMM models were fitted for 2, 3, and 4 classes, where the dependent variable corresponds to ART adherence. The random effects we included are the intercept and time changes, expressed through a longitudinal behavior. The fixed effects include only a third-grade polynomial of time. Patients' sex and age were initially considered, but there were no significant changes in the estimated adherence trajectories. The proposed model was chosen based on the AIC, BIC, entropy and relative entropy, shown in Table 1, the best results were found under LCMM with 2 latent classes of ART adherence, having the lowest BIC and the highest entropy. We used a grid of initial points, and the results show convergence.

[1] https://www.cdc.gov/mmwr/preview/mmwrhtml/rr5710a2.htm

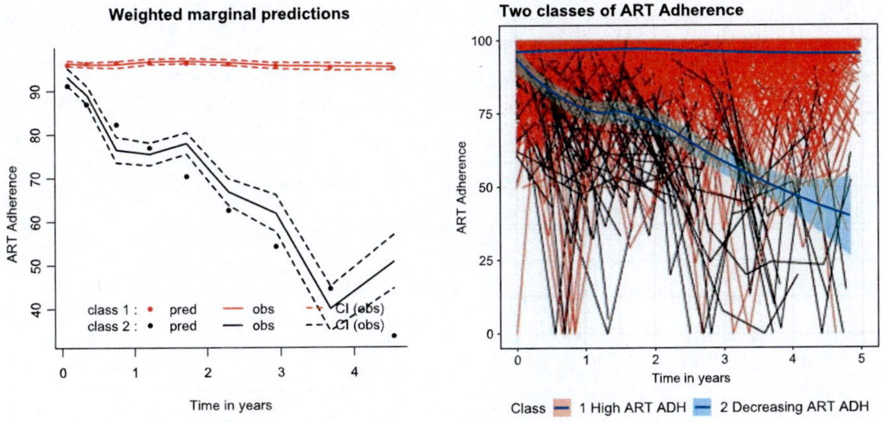

(a) (b)

Fig. 1 **a** Weighted marginal predictions for the ART adherence estimated by the LCMM with two classes. **b** Observed profiles of the two classes for the ART adherence; blue solid lines show the mean profiles computed with the loess smoothing method

The best model estimates two types of trajectories for the ART adherence with trajectories shown in Fig. 1. The first class has 924 (93.52% of the total population) patients who have high levels of ART adherence for a long time, in average 95%. The second class is a very small class of 64 (6.48% of the total population) patients whose adherence decreases along time. This class started with a 95% average as well, but decreased with time, reaching an average of 40% in the last observed time.

Table 2 shows the categorical variables describing the patient's characteristics per class of adherence. It summarizes the number of patients for the following variables; sex (male, female), previous AIDS defining event (conditions indicator previous to enrollment in care; no, yes), and CD4+T counts at the beginning of follow-up (in three categories: ≤ 200, 201–500, and >500 cells/mm^3). In order to explore if there are differences between classes, we used the Pearson's Chi-squared test with Yates' continuity correction. All p-values are greater than the significance level $\alpha = 0.05$, showing there are no differences between classes.

Table 3 summarizes continuous variables per class of patients, it shows the median and interquartile range for age, CD4/CD8 ratio, CD4+T count, and CD8+T count and ART adherence, both at the beginning and at the end of the follow-up. Differences in location for the distributions between classes of latent classes were tested using Wilcoxon's rank sum test with continuity correction. All p-values are greater than the significance level $\alpha = 0.05$, showing there are no differences between classes except for the ART adherence and time. In addition, the time at the end of follow-up, which varies for individuals, is summarized per class, people who have taken medications for a longer period (class 1, high levels of ART adherence) have more adherence in comparison to people who have taken medications for a shorter period (class 2, decreasing ART adherence).

Table 2 Patient's characteristics by class of ART adherence computed by LCMM (cases and percentage), and Pearson's Chi-squared test

Characteristics		Classes of ART ADH		Chi-squared test	
		1 High ($n = 924$)	**2** Decreasing ($n = 64$)	χ-squared	p-value
Sex	Male	838 (90.7%)	56 (87.5%)	0.3863	0.5342
	Female	86 (9.3%)	8 (12.5%)		
Previous AIDS	No	649 (70.2%)	52 (81.2%)	3.0075	0.0829
defining event[1]	Yes	275 (29.8%)	12 (18.8%)		
Baseline CD4+T	≤200	320 (34.6%)	21 (32.8%)	3.5071	0.1732
counts	201–500	443 (48.0%)	26 (40.6%)		
(cells/μL)[1]	>500	161 (17.4%)	17 (26.6%)		

[1]Condition previous to enrollment in care.

Table 3 Patient's characteristics by class of ART adherence computed by LCMM (median and interquartile range), and Wilcoxon's rank sum test

Characteristics		Classes of ART ADH		Wilcoxon test	
		1 High ($n = 924$)	**2** Decreasing ($n = 64$)	W	p-value
Age in years[2]		33 (28–41)	32 (28–7)	33590	0.0685
Time[3] at the end of follow-up		4.33 (1.96–4.75)	2.23 (1.37–3.46)	17780	<0.001
CD4/CD8	Beginning[2]	0.328 (0.188–0.530)	0.367 (0.210–0.577)	27107	0.2651
ratio	End[3]	0.647 (0.445–0.915)	0.602 (0.331–0.914)	32934	0.1275
CD4+T counts	Beginning[2]	287 (158–438)	322 (162–507)	26493	0.1637
	End[3]	486 (338–646)	461 (329–667)	30038	0.8318
CD8+T counts	Beginning[2]	845 (575–167)	921 (549–1241)	28028	0.4856
	End[3]	723 (550–971)	907 (501–1145)	25884	0.0953
Percentage of	Beginning[2]	100 (97.5–100)	97.8 (86.0–100)	36529	0.0006
ART Adherence	End[3]	98.9 (95.6–100)	52.2 (37.5–70.8)	57197	<0.001

[2]At the beginning of follow-up.
[3]At the end of follow-up (varies for individuals).

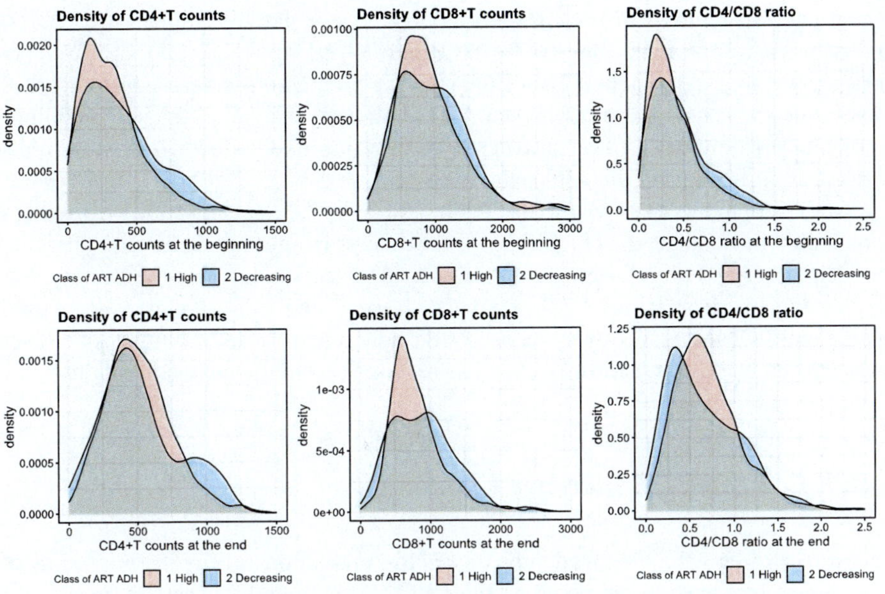

Fig. 2 Density of CD4+T counts, CD8+T counts and CD4/CD8 ratio by ART adherence latent class. Top: at the beginning of the follow-up. Bottom: at the end of the follow-up

4.2 *Latent Classes for Bivariate Response of CD4/CD8 Ratio and CD4+T Within Groups of Adherence*

CD4/CD8 ratio and CD4+T counts have been used as biomarkers of disease stage in PWH (p.e. [19]). Despite their relationship, they are usually studied sometimes using linear mixed models (LMM), but not jointly in a bivariate response model (p.e. [20]). We notice that a CD4/CD8 ratio equal to 1 could be obtained from CD4+T counts of 900 and CD8+T counts of 900, but also from 200, or any other quantity with the same CD4+T and CD8+T values. However, having CD4+T of 900 is a normal patient's condition, but having CD4+T of 200 indicates a high risk of opportunistic infections.

Figure 2 shows the density of these variables per ART adherence class at the beginning and end of the follow-up, computed by kernel density estimates. We believe there are different features of patients within each class and a joint analysis could provide a partition that could help to identify characteristics linked to consistently low ratios. In this section, we show the joint analysis of the CD4/CD8 ratio and the CD4+T counts, using the latent process mixed model for these multivariate longitudinal markers (multLCMM) [7].

The dependent variable is the bivariate vector of biomarkers CD4/CD8 ratio and CD4+T in logarithm scale, i.e., (log CD4/CD8 ratio, log CD4+T). Random effects for the intercept and time were included, using a Brownian motion for its variance–

covariance structure. The independent covariates are the ART adherence (ADH) measured in a continuous range of 0–100% and a 4-order polynomial for the time. In order to explore other factors associated with the biomarkers, we have also included sex (male vs. female), age in years at ART initiation, and AIDS-related conditions indicator previous to enrollment in care (yes or no) as fixed effects. Lastly, the class effects are the time and the ART adherence.

After fitting the LCMM for the ART adherence in Sect. 4.1, the two classes obtained will be labeled as high (class 1) and decreasing (class 2) levels. The joint models described above were fitted for each ART adherence class separately. For both (high and decreasing) ART adherence classes 3 and 2, latent classes were used in the multLCMM, respectively. In each ART adherence class, the number of classes was chosen by BIC and entropy. Results are described in the next subsections.

High Levels ART Adherence Class

Figure 3 displays the weighted subject-specific predictions of the latent process of the three latent classes estimated by the multLCMM, and the observed profiles for CD4/CD8 ratio, CD4+T counts and CD8+T counts computed by the loess smoothing method. Estimated parameters are summarized in Table 4. From the negative parameter estimates for age, sex, and AIDS, we observe that older male patients with AIDS show the least increase of the joint values of the response.

Table 5 shows patient's characteristics for the subclasses in the high levels of ART adherence class (cases, or median and interquartile range). In order to test differences by latent classes, we used the Pearson's Chi-squared test with Yates' continuity correction (for categorical variables) and Kruskal-Wallis test (for continuous variables). Results show there are no significant differences in sex, age, and CD8+T at the beginning of follow-up. There are significant differences for AIDS, CD4+T counts, and CD4/CD8 ratio. This means that latent classes are characterized mainly by the AIDS condition and levels of CD4+T counts and CD4/CD8 ratio at baseline.

From Fig. 3, we note that class 1 encompasses most of the patients; the CD4/CD8 ratio has different trajectory behavior for classes 2 and 3; class 2 has almost parallel trajectories to those of class 1, the latter in a lower level. Class 3 has a constant trajectory along time. Since these are all patients who have high levels of ART adherence, this could imply that the patient's condition is not only determined by adherence to treatment. There are also patients whose specific characteristics will lead to poor CD4+T and CD4/CD8 ratio conditions, and this condition will prevail, only among those with the worst initial condition.

Table 4 High levels of ART adherence class: estimates of the multLCMM with bivariate response (log CD4/CD8 ratio, log CD4+T)

Variable	Estimate	SE	Wald	*p*-value
Fixed effects in the longitudinal model				
ART ADH class1	0.00051	0.00053	0.948	0.34303
ART ADH class2	0.00710	0.00228	3.109	0.00187
ART ADH class3	–0.03314	0.00474	–6.989	0.00000
AIDS: Yes	–0.20878	0.05515	–3.786	0.00015
Age	–0.00904	0.00253	–3.786	0.00035
Sex: Female	0.24589	0.08206	2.996	0.00273
Parameters of the link functions H_d in (1)				
log CD4/CD8 ratio -Linear η_1	–0.89597	0.07395	–12.116	0.00000
log CD4/CD8 ratio -Linear η_2	0.66828	0.02058	32.468	0.00000
log CD4 -Linear η_1	5.75555	0.05869	98.070	0.00000
log CD4 -Linear η_2	0.52911	0.01694	31.231	0.00000

Decreasing ART Adherence Class

Figure 4 displays the weighted subject-specific predictions of the latent process of the two latent classes estimated by the multLCMM, and the observed profiles for CD4/CD8 ratio, CD4+T counts, and CD8+T counts, computed by the loess smoothing method. Estimated parameters are summarized in Table 6. In this case, from the negative parameter estimates for age, sex, and AIDS, we observed that the lower joint increase would again correspond to older male patients with AIDS.

Table 7 shows patient's characteristics for classes of patients within the decreasing ART adherence class (cases, or median and interquartile range). In order to test differences by latent classes, we use the Pearson's Chi-squared test with Yates' continuity correction (for categorical variables) and Wilcoxon's rank sum test (for continuous variables).

The results in the decreasing ART adherence class analysis show there are no significant differences for sex, age, AIDS, and CD8+T, however, differences for CD4+T counts and CD4/CD8 ratio are significant. This suggests latent classes are characterized by the CD4+T counts and CD4/CD8 ratio at baseline, but it is about the decreasing ART adherence class. Note that once again patients who started with low levels of CD4+T counts and CD4/CD8 ratio will continue in similar conditions over time, regardless of their lower levels of ART adherence.

Table 5 High levels of ART adherence class: characteristics (cases, or median and interquartile range) of patients from LCMM with bivariate response (CD4/CD8 ratio, CD4+T), Pearson's Chi-squared test, and Kruskal-Wallis test

High ART Adherence		Classes from multLCMM			Chi-squared test	
		1	**2**	**3**	χ-squared	p-value
Characteristics		(n = 861)	(n = 16)	(n = 47)		
Sex	Male	783	13	42	1.851	0.3962
	Female	78	3	5		
Previous AIDS	No	612	6	31	8.907	0.0116
defining event[1]	Yes	249	10	16		
Baseline CD4+T	≤200	271	15	34	58.088	<0.001
counts	201–500	432	1	10		
(cells/μL)[1]	>500	158	0	3		
Hight ART Adherence		Classes from multLCMM			Kruskal-Wallis test	
Characteristics		**1**	**2**	**3**	χ	p-value
Age in years[2]		33 (28–41)	37 (34–46)	36 (30–45)	5.540	0.0626
CD4/CD8 ratio	Beginning[2]	0.341 (0.203–0.539)	0.074 (0.045–0.094)	0.188 (0.127–0.288)	58.974	<0.001
	End[3]	0.676 (0.492–0.931)	0.229 (0.095–0.329)	0.280 (0.188–0.323)	128.460	<0.001
CD4+T counts	Beginning[2]	301 (169–446)	37 (19–89)	158 (112–212)	53.134	<0.001
	End[3]	502 (361–658)	169 (87–314)	242 (140–327)	91.362	<0.001
CD8+T counts	Beginning[2]	845 (584–1170)	734 (460–931)	896 (540–1262)	2.529	0.2824
	End[3]	717 (549–952)	885 (457–1254)	857 (592–1324)	5.956	0.0509

[1]Condition previous to enrollment in care.
[2]At the beginning of follow-up.
[3]At the end of follow-up (varies for individuals).
[3]At the end of follow-up (varies for individuals).

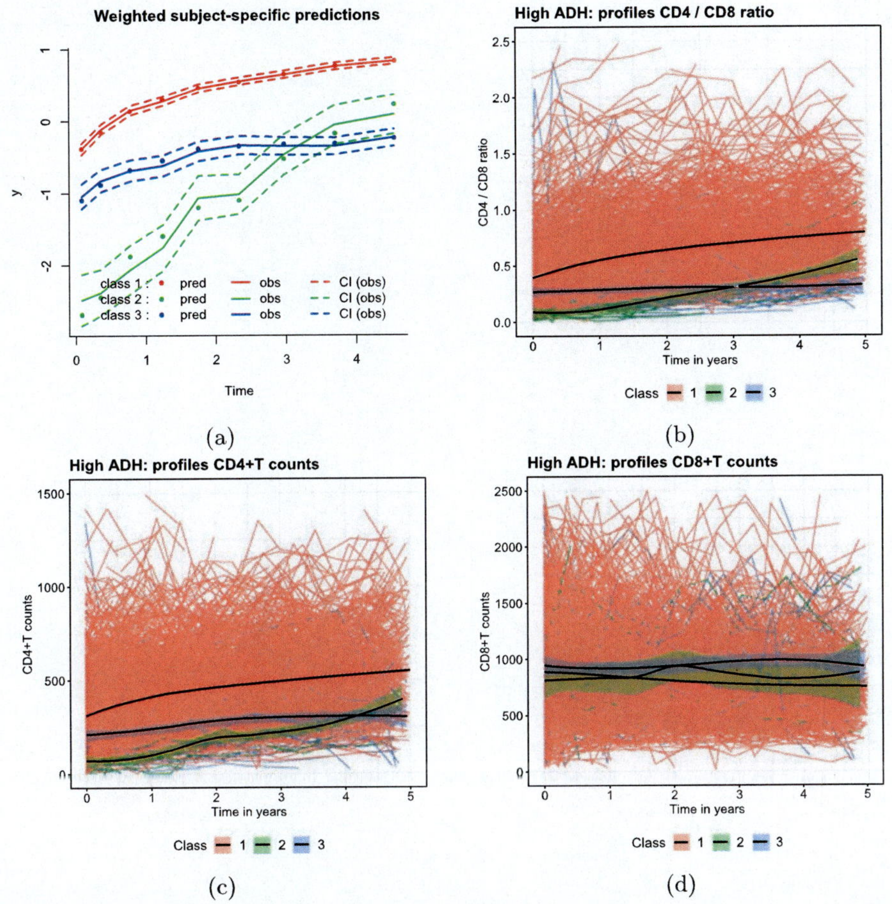

Fig. 3 Hight levels of ART adherence class: three classes from the multLCMM with bivariate response (log CD4/CD8 ratio, log CD4+T). **a** Weighted subject-specific predictions estimated by the multLCMM with three classes. Observed profiles of the three classes for (**b**) CD4/CD8 ratio, (**c**) CD4+T counts, and (**d**) CD8+T counts; solid lines show the mean profiles computed with the loess smoothing method

5 Discussion and Conclusions

Using LCMM to model ART adherence, we found two groups of trajectories, within each group a joint model for CD4+T counts and CD4/CD8 ratio distinguished heterogeneous subpopulations.

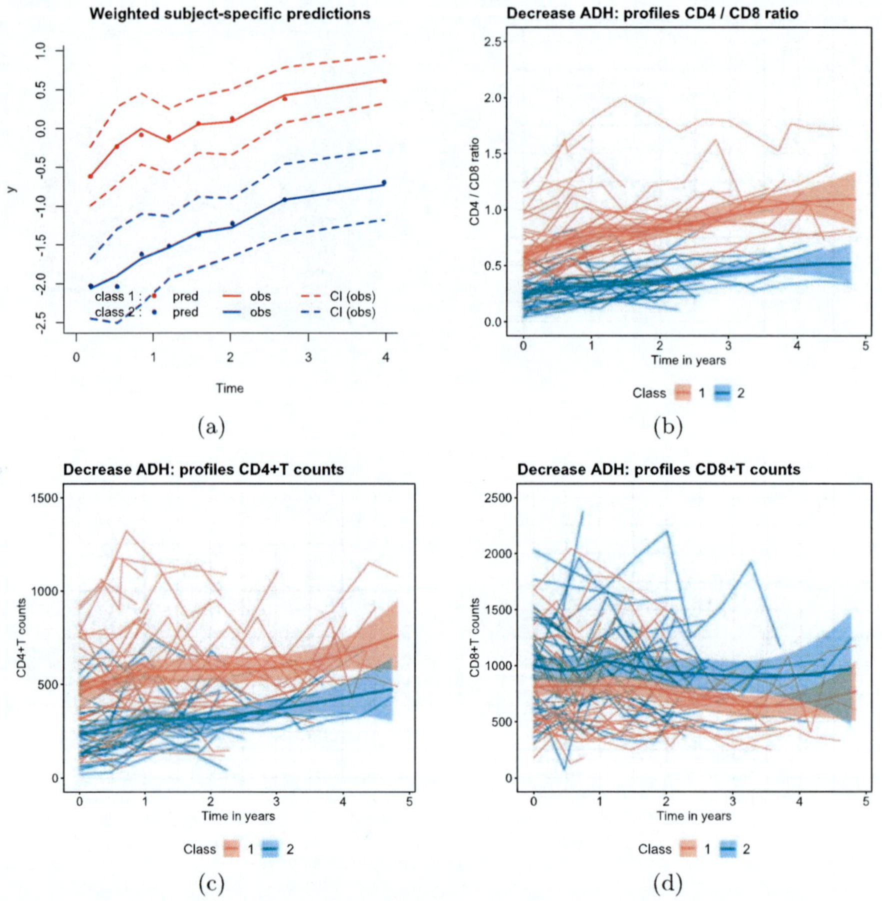

Fig. 4 Decreasing ART adherence class: two classes from a multLCMM with bivariate response (log CD4/CD8 ratio, log CD4+T). **a** Weighted subject-specific predictions estimated by the multLCMM with two classes. Observed profiles of the three classes for (**b**) CD4/CD8 ratio, (**c**) CD4+T counts and (**d**) CD8+T counts; solid lines show the mean profiles computed with the loess smoothing method

The majority of PWH receiving care at the HIV/AIDS Clinic at the INCMNSZ initiate and maintain a good level of ART adherence throughout time. However, there is a small subset whose adherence decreases to levels lower than 50% (class 2 Decreasing ART adherence in Sect. 4.1). These patients have a lower proportion of AIDS at enrollment in care (see Table 2) and show a lower increase in their CD4+T counts and CD4/CD8 ratio between the beginning and the end of the follow-up (see Table 3). Within this decreasing ART adherence class, we found a class of patients

Table 6 Decreasing ART adherence class: estimates of the multLCMM with bivariate response (log CD4/CD8 ratio, log CD4+T)

Variable	Estimate	SE	Wald	p-value
Fixed effects in the longitudinal model				
ADH class1	0.00051	0.00125	0.408	0.68305
ADH class2	0.00365	0.00145	2.513	0.01197
AIDS: Yes	–1.32212	0.20963	–6.307	0.00000
Age	–0.01434	0.00676	–2.122	0.03382
Sex: Female	0.60809	0.19330	3.146	0.00166
Parameters of the link functions H_d in (1)				
log CD4/CD8 ratio -Linear η_1	–0.34856	0.17443	–1.998	0.04569
log CD4/CD8 ratio -Linear η_2	0.57959	0.06167	9.399	0.00000
log CD4 -Linear η_1	6.20719	0.13867	44.762	0.00000
log CD4 -Linear η_2	0.45373	0.05166	8.782	0.00000

who started with low CD4+T counts and CD4/CD8 ratio lower than 0.5 (see Fig. 4 and Table 7); for these patients, the initial ART adherence level shows a higher effect on the bivariate response in the joint model (see Table 6). Therefore, age, sex, AIDS, and initial CD4+T counts seem to be crucial to maintain good CD4/CD8 ratios for those who start with adverse conditions, good ART adherence will provide a desirable positive effect in CD4+T counts and CD4/CD8 ratio. Additionally, further study of these classes could include specific analysis for clinical outcomes.

Latent class mixed models enable a useful interpretation of longitudinal heterogeneity in the population of interest. Identifying and characterizing different families of trajectories of a specific outcome, could focus the results of a model, allowing to detect covariates related to the difference and getting better and more specific answers to research questions. The mean profiles of the classes are defined according to time and covariates through latent class-specific mixed models. In this study, both fixed and random effects can be class specific. This is an advantage in comparison to the standard linear mixed models where the population is considered homogeneous and only one mean profile of response is needed. Moreover, the lcmm allows the study of multivariate longitudinal markers instead of a unique one.

Table 7 Decreasing ART adherence class: characteristics (cases, or median and interquartile range) of patients from multLCMM with bivariate response (log CD4/CD8 ratio, log CD4+T), Pearson's Chi-squared test, and Wilcoxon rank sum test

Decrease ART Adherence		Classes from multLCMM		Chi-squared test	
		1	**2**	χ-squared	p-value
Characteristics		($n = 36$)	($n = 28$)		
Sex	Male	32	24	0.0001	1
	Female	4	4		
Previous AIDS	No	28	24	0.2344	0.6283
defining event[1]	Yes	8	4		
Baseline CD4+T	≤200	7	14	11.61	0.0030
counts	201–500	14	12		
(cells/μL)[1]	>500	15	2		
Decrease ART Adherence		Classes from multLCMM		Wilcoxon test	
Characteristics		**1**	**2**	W	p-value
Age in years[2]		31 (27–35)	32 (28–38)	472	0.6720
CD4/CD8 ratio	Beginning[2]	0.560 (0.394–0.808)	0.218 (0.130–0.313)	875	<0.0001
	End[3]	0.847 (0.612–1.143)	0.347 (0.238–0.406)	869	<0.0001
CD4+T counts	Beginning[2]	478 (293–711)	201 (127–344)	773	0.0003
	End[3]	567 (441–886)	345 (291–460)	785	0.0002
CD8+T counts	Beginning[2]	799 (501–1156)	1071 (687–1429)	373	0.0774
	End[3]	721 (450–1083)	990 (735–1213)	366	0.0628

[1]Condition previous to enrollment in care.
[2]At the beginning of follow-up.
[3]At the end of follow-up (varies for individuals).

Acknowledgements This work was supported by the NIH-funded Caribbean, Central and South America network for HIV epidemiology (CCASAnet), a member cohort of the International Epidemiologic Databases to Evaluate AIDS (leDEA) (U01AI069923). The content is solely the responsibility of the authors and does not necessarily represent the official views of the National Institutes of Health. This award is funded by the following institutes: National Institute Of Allergy And Infectious Diseases (NIAID), Eunice Kennedy Shriver National Institute Of Child Health & Human Development (NICHD), National Heart, Lung, And Blood Institute (NHLBI), National Institute Of Diabetes And Digestive And Kidney Diseases (NIDDK), National Institute On Drug Abuse

(NIDA), National Institute On Alcohol Abuse And Alcoholism (NIAAA), Fogarty International Center (FIC), and National Cancer Institute (NCI). LN and RFG have been in part supported by UNAM-DGAPA-PAPIIT, Mexico (Project IN118720). Thanks to Dra. Brenda Crabtree-Ramírez and Dr. Juan G. Sierra-Madero from Departamento de Infectología, Instituto Nacional de Ciencias Médicas y Nutrición Salvador Zubirán, Mexico City, Mexico, for data, clinical information and discussion. We are also grateful to two anonymous reviewers who provided insightful comments and suggestions that have greatly improved both the readability and the content of this article.

References

1. Calmy, A., Tovar Sanchez, T., Kouanfack, C., et al.: Dolutegravir-based and low-dose efavirenz-based regimen for the initial treatment of HIV-1 infection (NAMSAL): Week 96 results from a two-group, multicenter, randomized, open label, phase 3 non-inferiority trial in Cameroon. Lancet HIV 7(10), e677–e687 (2020)
2. Caro-Vega, Y., Belaunzarán-Zamudio, P.F., Crabtree-Ramírez, B.E., et al.: Durability of efavirenz compared with boosted protease inhibitor-based regimens in antiretroviral-naïve patients in the Caribbean and Central and South America. Open Forum Infect. Dis. 5(3), ofy004 (2018)
3. Wohl, D.A., Yazdanpanah, Y., Baumgarten, A., et al.: Bictegravir combined with emtricitabine and tenofovir alafenamide versus dolutegravir, abacavir, and lamivudine for initial treatment of HIV-1 infection: week 96 results from a randomized, double-blind, multicenter, phase 3, non-inferiority trial. Lancet HIV 6(6), e355–e363 (2019)
4. Castillo-Mancilla, J.R., Brown, T.T., Erlandson, K.M., et al.: Suboptimal adherence to combination antiretroviral therapy is associated with higher levels of inflammation despite HIV suppression. Clin. Infect. Dis.: Offic. Publ. Infect. Dis. Soc. Am. 63(12), 1661–1667 (2016)
5. Castillo-Mancilla, J.R., Brown, T.T., Palella, F.J.J., et al.: Partial normalization of biomarkers of inflammation and immune activation among virally suppressed men with HIV infection and high ART adherence. Open Forum Infect. Dis. 7(4), ofaa099 (2020)
6. Castillo-Mancilla, J.R., Phillips, A.N., Neaton, J.D., et al.: Incomplete ART adherence is associated with higher inflammation in individuals who achieved virologic suppression in the START study. J. Int. AIDS Soc. 22(6), e25,297 (2019)
7. Proust-Lima, C., Philipps, V., Liquet, B.: Estimation of extended mixed models using latent classes and latent processes: The R package LCMM. J. Stat. Softw. 78(2), 1–56 (2017)
8. Muthén, B.: Latent variable analysis: Growth mixture modeling and related techniques for longitudinal data. In: Kaplan, D. (ed.) Handbook of Quantitative Methodology for the Social Sciences, pp. 345–368. Sage, Newbury Parc, CA (2004)
9. Muthén, B.O.: Beyond SEM: General latent variable modeling. Behaviormetrika 29(1), 81–117 (2002)
10. Bruno, G., Saracino, A., Monno, L., Angarano, G.: The revival of an old marker: CD4/CD8 ratio. AIDS Rev. 19(2), 81–88 (2017)
11. Belaunzarán-Zamudio, P.F., Naranjo, L., Caro-Vega, Y., Castillo-Mancilla, J.R., Camiro-Zuñiga, A., Fuentes-García, R., Crabtree-Ramírez, B.E., Sierra-Madero, J.G.: Incomplete ART adherence is associated with lower CD4-CD8 ratio in virally suppressed patients with HIV-infection in Mexico. Submitted (2021)
12. Sierra-Madero, J.G., Belaunzarán-Zamudio, P.F., Crabtree-Ramírez, B., Magis-Rodriguez, C.: Mexico's fragmented health system as a barrier to HIV care. Lancet HIV 6(2), e74–e75 (2019)
13. Belaunzarán-Zamudio, P., Caro-Vega, Y.N., Shepherd, B.E., et al.: Monitoring of HIV treatment in seven countries in the WHO Region of the Americas. Bull. World Health Organ. 93(8), 529–539 (2015)
14. McMahon, J.H., Jordan, M.R., Kelley, K., et al.: Pharmacy adherence measures to assess adherence to antiretroviral therapy: review of the literature and implications for treatment monitoring. Clin. Infect. Dis. 52(4), 493–506 (2011)

15. Boussari, O., Subtil, F., Genolini, C., et al.: Impact of variability in adherence to HIV antiretroviral therapy on the immunovirological response and mortality. BMC Med. Res. Methodol. 15, 10 (2015)
16. Akaike, H.: Information theory and an extension of the maximum likelihood principle. In: B.N. Petrov, F. Csaki (eds.) Proceedings of 2nd International Symposium on Information Theory, pp. 267–281. Académiai Kiado, Budapest, Hungary (1973)
17. Schwarz, G.: Estimating the dimension of a model. Ann. Stat. 6(2), 461–464 (1978)
18. Celeux, G., Soromenho, G.: An entropy criterion for assessing the number of clusters in a mixture model. J. Classif. 13(2), 195–212 (1996)
19. Sauter, R., Huang, R., Ledergerber, B., et al.: CD4/CD8 ratio and CD8 counts predict CD4 response in HIV-1-infected drug naive and in patients on cART. Medicine 95(42), e5094 (2016)
20. Wang, W.L., Fan, T.H.: Bayesian analysis of multivariate t linear mixed models using a combination of IBF and Gibbs samplers. J. Multivar. Anal. 105(1), 300–310 (2012)

A Dynamic Model for Analyzing the Public Health Policy of the Mexican Government During the COVID-19 Pandemic

E. Iván Guerrero-Flores, J. Héctor Morales-Bárcenas, and Gabriel Núñez-Antonio

1 Introduction

The coronavirus disease (COVID-19) pandemic has been a significant health threat to the whole of humankind. According to the World Health Organization (WHO) [28], there have been more than 223 million confirmed cases of COVID-19 worldwide, including more than 4.6 million deaths. Despite the progress in scientific knowledge in human behavior, epidemiology, virus ecology, and evolution, there are still missing factors to fully understand the progression of the infectious disease in human populations with the coronavirus family [19]. On the other hand, most governments adopted vaccine schemes and mobility restrictions to mitigate the spread of the SARS-CoV-2 and its recent variants. However, these restrictions have a significant economic impact, and we are far from foreseeing the consequences. Our goal in this paper is to assess the public health policy on the population mobility of the Mexican government quantitatively from 2019 to 2021. We want to contribute to the ongoing studies aiming to improve public health policies [24]. In Mexico, from January 2020 until September 2021, there have been 3,465,171 confirmed cases of COVID-19 with 265,420 deaths, reported to WHO [28]. We carry out database analysis support on Mexico's National Council for Science and Technology (CONACyT) and Secretariat of Health [8]. The availability of these data drives us to choose an appropriate dynamic system to describe the pandemic (see Fig. 1). We represent the daily and weekly data points and their respective cumulative number of cases and deaths.

E. I. Guerrero-Flores · J. H. Morales-Bárcenas · G. Núñez-Antonio (✉)
Departamento de Matemáticas, Universidad Autónoma Metropolitana Unidad Iztapalapa (UAM -I), Av. San Rafael Atlixco 186, Iztapalapa, CP 09340 Mexico City, Mexico
e-mail: gab.nuneza@gmail.com

J. H. Morales-Bárcenas
e-mail: jhmb@xanum.uam.mx

I. Antoniano-Villalobos et al. (eds.), *Interdisciplinary Statistics in Mexico*,
Springer Proceedings in Mathematics & Statistics 397,
https://doi.org/10.1007/978-3-031-12778-6_7

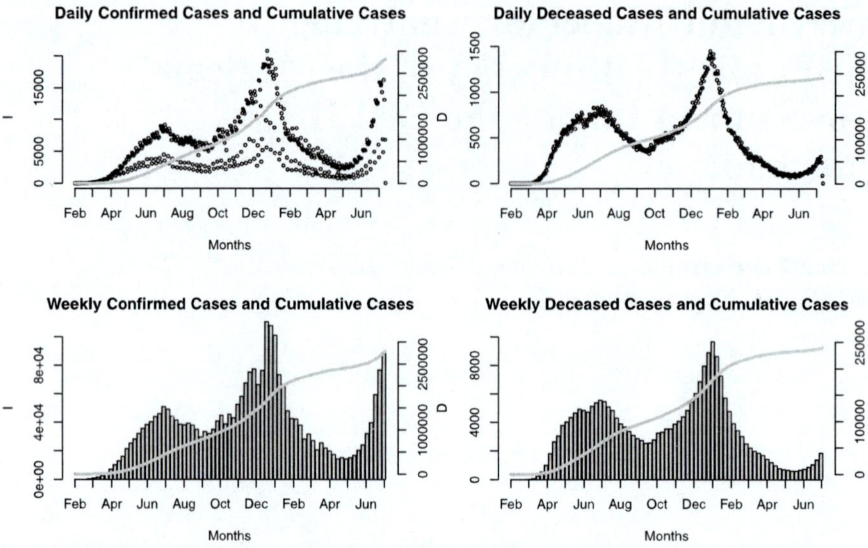

Fig. 1 Daily and weekly confirmed and deceased cases during the COVID-19 Pandemic in Mexico. Faint gray lines represent the cumulative confirmed and deceased cases

We observe up to three waves[1] on the data (the third one is under development). This work will associate wave dates with parametric structures describing the Mexican government's set of public health policies. We will manage the weekly registered data to better explain the pandemic trend. We use the mechanistic SIRD model [10] to describe the data trends in confirmed and dead disease cases. Much of the research on the subject has focused on determining the unknown parameters of these kinds of ordinary differential equations (ODE) systems but less on estimating their posterior distribution under a Bayesian framework [12, 13, 23]. We performed Bayesian inference using a Markov Chain Monte Carlo (MCMC) method on the proposed model in this last direction. Following [17], we choose a time delay exponential function to set the transmission rate changes into the model.

2 Related Work

Since 2019, when the COVID-19 pandemic began in China, many papers have appeared explaining the pandemic using the mechanistic Susceptible-Infected-Recovered (SIR) model variations [2, 22]. Sometimes these models inaccurately assume a homogeneous population mixing, excluding heterogeneity arising from age, space, or behavioral aspects [9, 15]. Despite these shortcomings, most of these studies seek to predict the course of the pandemic using the point-estimator approach

[1] We understand a graph that tracks an epidemic's evolution over time for an *epidemic wave*, implying, for instance, a rising number of infected individuals, a more or less defined peak, and a decline.

for model parameters, placing out the uncertainty features of the observed data [1]. We select a Bayesian approach [11, 13, 21] to estimate conditional probabilities for model parameters, calculate credibility regions over our predictions' primary trend, and do correlation analysis to get a deep understanding of the model parameters subject to the data [12]. This approach will allow us to compare similar analyses from different databases and models. Previous studies also assume that the transmission rate β is constant (see Eq. (2)). We relax this restriction to cover the general case of time-dependent parameters. The time-dependent $\beta(t)$ will reflect the adjusted public health measures during the pandemic [5]. Following [17], we introduce an exponential transmission rate $\beta(t)$ to model the changing course of the pandemic dynamics due to the government mobility measures:

$$\beta(t) = \begin{cases} \beta_0, & \text{if} \quad t < t_0, \\ \beta_0 \exp\left(-\mu_0(t - t_0)\right), & \text{if} \quad t \geq t_0. \end{cases} \tag{1}$$

The parameter μ_0 measures the impact of the government intervention to lower the transmission rate. The smaller the value (μ_0), the lesser effectiveness of the intervention ($\beta(t)$). The parameter β_0 is called the transmission rate constant and informs the average rate of new infections of individuals. The higher the value, the bigger the force of infection [15]. The "Jornada de sana distancia" (Government Social Distant Campaign) took effect in Mexico on the fourth week after the first confirmed case in February 2020; in this case, $t_0 = 4$ in Eq. (1). Up to date, two more governmental interventions fixed the other two times of the intervention (see below Eq. (3)).

3 Methods

There is a rich tradition in mathematical epidemiology to use models to match the pandemic dynamics progression [3, 5, 14]. Specifically, we use the SIRD model to investigate the epidemic waves that travel through an entire population and interpret model parameters guided by data. SIRD model is a compartmental model that distributes the host population into susceptibles, infectious, recovered, and dead classes and tracks changes in the number of the population within each category. We adjust the model to estimate the costs and benefits of intervention strategies to limit disease risks, such as quarantine.

3.1 SIRD Model

We introduce the recovery class $R(t)$ and the deceased class $D(t)$ as follows. Individuals leave the susceptible class $S(t)$ through infection after encountering infected hots $I(t)$ at rate $\beta SI/N$ (mass action law), where N is the total population size. Infected hosts grow to encounter susceptibles and drop their class through natural recovery rate γI and disease-induced mortality rate αI. The mathematical SIRD model is then the following system of the ordinary differential equations (in the frequency dependent formulation (see [15])):

$$S' = -\beta(t)SI/N, \tag{2a}$$

$$I' = \beta(t)SI/N - \gamma I - \alpha I, \tag{2b}$$

$$R' = \gamma I, \tag{2c}$$

$$D' = \alpha I, \tag{2d}$$

where prime $'$ means time derivative, the parameter $\beta(t)$ is the infection rate at time t, γ the recovery rate and α the mortality rate from COVID-19. From Eq. (2), we observe that the total population size N is a conserved quantity (constant), such that, $N = S(t) + I(t) + R(t) + D(t)$. SIRD model is an initial value problem for a non-linear coupled ordinary differential equations, then we impose the initial values $S(0) = S_0 \geq 0$, $I(0) = I_0 \geq 0$, $R(0) = R_0 \geq 0$, and $D(0) = D_0 \geq 0$. The estimation of the *force of infection*, λ, is another critical parameter to assess the progression of the pandemic. We define $\lambda = \beta I/N$ as the production rate (*per capital*) of new infections [15, 27]. Many authors have indeed analyzed the SIRD model in the past. For instance, the theoretical aspects of the model, such as the phase space approaches, will not be considered here (see [3, 9, 14, 15]). As we are going to divide the time into three COVID-19 waves, then we are going to have α_j and $\beta_j(t)$ for each wave $j = 1, 2, 3$. We will use a fixed value of γ for all of the three waves.

3.2 Transmission Rate Model

We analyze each COVID-19 wave breaking the entire observational time into three waves: $j = 1, 2,$ and 3. We define the general infection rate structure through Eq. (3):

$$\beta_j(t) = \begin{cases} \beta_j, & \text{if } t < t_j, \\ \beta_j \exp\left(-\mu_j\left(t - t_j\right)\right), & \text{if } t \geq t_j. \end{cases} \tag{3}$$

We display wave dates, t_j, in Table 1 that mark the exponential decaying of the transmission rate. We hypothesize that these points agree with the government intervention, namely, canceling massive events, home confinement, wearing masks, and getting social distance.

The beginning of the Social Distant Campaign and the cancelation of face-to-face school classes marked the first intervention date, $t_1 = 4$ (fourth week), around March 17th to 24th, 2020. Mass cancelations and the epidemiological color-coded map changed to prevention (orange) marked the second intervention date, $t_2 = 41$ (forty-

Table 1 Indexing for the model parameters

Wave j	Dates for each wave	Intervention times t_j	Time range (weeks)
1	Feb 18th, 2020–Sep 22th, 2020	$t_1 = 4$	$1 \leq t \leq 31$
2	Sep 22th, 2020–May 17th, 2021	$t_2 = 41$	$32 \leq t \leq 65$
3	May 17th, 2021–Jul 26th, 2021	$t_3 = 71$	$66 \leq t \leq 75$

first week), around December 1st to 8th, 2020. During the last intervention date, $t_3 = 71$ (seventy-first week), between June 29th and July 6th, 2021, the Secretariat of Health of Mexico reported fast-increasing infections. As a consequence, restrictions started to come back. A few weeks back, the COVID-19 cases had been escalated in the states of Baja California and Yucatan and on Mexico's northern border.

3.3 Recovery Rate Estimation

We mention that the only available data are those about classes I and D. However, the database reports the latest cumulative records from the recovered class R. We took two consecutive weekly records from this database to estimate the recovery rate: $\gamma - \Delta R/I \approx 0.94$ [16].

3.4 Bayesian Inference

Consider n data points for both classes I and D, which represent the weekly confirmed cases and the accumulated deaths on the i-th week. We propose the following dynamic model

$$y_{ij} = x_j(t_i|\boldsymbol{\theta}, \boldsymbol{x}_0) + \varepsilon_j, \tag{4}$$

where $x_j(t_i|\boldsymbol{\theta}, \boldsymbol{x}_0)$ are the solutions of the Eq. (2) [20] for I and D ($j = 1, 2$) at the time t_i, ε_j is the noise error for the class j, with $\varepsilon_j \sim N(0, \sigma_j^2)$. The vector $\boldsymbol{\theta} = (\beta_1, \beta_2, \beta_3, \mu_1, \mu_2, \mu_3, \alpha_1, \alpha_2, \alpha_3, \sigma_1^2, \sigma_2^2)^T$ comprises all the system parameters, and $\boldsymbol{x}_0$ are the initial conditions of the system. Therefore, the likelihood function is given by

$$p(\boldsymbol{y}|\boldsymbol{\theta}) = \prod_{j=1}^{2}\prod_{i=1}^{n} N(y_{ij}|x_j(t_i|\boldsymbol{\theta}, \boldsymbol{x}_0), \sigma_j^2). \tag{5}$$

Table 2 Log-normal (LN) prior specifications for each pandemic wave $j = 1, 2, 3$

Parameter	Prior distribution
β_j	$LN(0, 1)$
μ_j	$LN(0, 1)$
α_j	$LN(0, 1)$
σ_1^2	$LN(18, 5)$
σ_2^2	$LN(18, 5)$

Our starting point for Bayesian inference is the posterior density $p(\boldsymbol{\theta}|\mathbf{y})$, which is given up to a constant of proportionality by

$$p(\boldsymbol{\theta}|\mathbf{y}) \propto p(\mathbf{y}|\boldsymbol{\theta})p(\boldsymbol{\theta}). \tag{6}$$

Inference on this kind of dynamic model is nontrivial. For instance, the nonlinearity of the ODE system can lead to multiple modes. Then, purely local optimization algorithms are unlikely to get a global fit. On the other hand, the posterior distribution does not have a closed form. Thus, we require methods such as MCMC for sampling from the posterior distribution. However, the selection of an MCMC method is nontrivial too. The computational cost can be prohibitive, as, on every iteration, the algorithm will require numerical integration to get a solution for the ODE system. We choose and implement the Robust Adaptive Metropolis (RAM) [25, 26]. The RAM method is an adaptive MCMC method to learn the best parameter values [4, 7]. The algorithm "estimates the shape of the target distribution and simultaneously coercing the acceptance rate. The adaptation rule is computationally simple, adding no extra cost compared with the Adaptive Metropolis algorithm" [26].

Before the sampling computing, we need to define the prior distribution for the model parameters. Nevertheless, the prior selection is a challenge because the COVID-19 is a new disease, and we still do not have enough information on its many statistical features. We then use a prior specification defined by marginal Log-Normal (LN) distributions for each parameter in the model (see Table 2).

4 Results

We implement the stated methods to analyze the COVID-19 open database from CONACyT (https://datos.covid-19.conacyt.mx/) and the Secretariat of Health of Mexico [8].

We took the initial values for our ODE system $S_0 = 127{,}792{,}286$, $I_0 = 2$, $R_0 = 0$, and $D_0 = 0$. We use priors defined in Table 2. We integrate the system with the fourth-order Runge-Kutta method [6]. In order to reduce the computational cost, we parallelize programming in R. The time computing was, on average, 1.5 h in

Table 3 Point estimations and 95% HPD credible regions

Parameter	Mean	Median	HPD regions
β_1	1.945	1.947	(1.9273 ; 1.9603)
β_2	1.195	1.196	(1.1748 ; 1.2123)
β_3	1.411	1.414	(1.3448 ; 1.4741)
μ_1	0.03402	0.03404	(0.0322 ; 0.0357)
μ_2	0.02276	0.02278	(0.0199 ; 0.0251)
μ_3	0.06652	0.06336	(0.0252 ; 0.1142)
α_1	0.1372	0.1369	(0.1213 ; 0.1543)
α_2	0.08989	0.08983	(0.0835 ; 0.0964)
α_3	0.044	0.04335	(0.0252 ; 0.0662)
σ_1	10,284	10,201	(8,436 ; 12,051)
σ_2	4,475	4,432	(3,649 ; 5,247)

a standard 64-bit computer with a Ryzen 7 3750H CPU and 16 GB of RAM. We applied the algorithm previously described to get a sample of size 1,200 from the posterior distribution of parameter vector $\boldsymbol{\theta}$. Convergence diagnostics led us to stop the process after 137,500 iterations by core (having eight cores), dropping the first 100,000 as a burn-up. From the remaining 37,500 iterations, we kept one observation every 250 iterations by core, joining our sample collected by each core as part of the final sample.

The posterior marginal distributions allow us to understand the pandemic's evolution. We show the marginal distribution outcomes for each parameter in Fig. 2. A decisive observation is that these distributions were time-shifting. The first row of Fig. 2 shows that the initial α distribution (mortality rate) shifted to the left (lower values), reducing its density from 15 to 5% toward the third wave. The second row shows that β (transmission rate) shifted over the three waves transition. In the third row, we must remark that the μ distributions overlapped. The initial μ distribution (first wave) shifted to the left a little on the second and third waves, increasing their density and reducing their dispersion. The distribution's support also shifted showing that the government intervention helped to decrease the transmission rate. Finally, in the fourth row, we show the variance σ_1^2 and σ_2^2 over the predictive trends of the classes I and D, respectively. Notably, the variance σ_1^2 for I is more prominent than σ_2^2 for D.

The Bayesian inference is a powerful method in applied modeling that allows us to calculate High Posterior Density (HPD) credible regions for the parameters over simple point estimations [21]. Table 3 shows these 95% credible regions for each component of the vector parameter $\boldsymbol{\theta}$. Combining sampling from the posterior distribution of $\boldsymbol{\theta}$ and numerically solving the ODE system allow us to sample from the posterior predictive distribution and obtain the credible regions at each point in time (see Fig. 3).

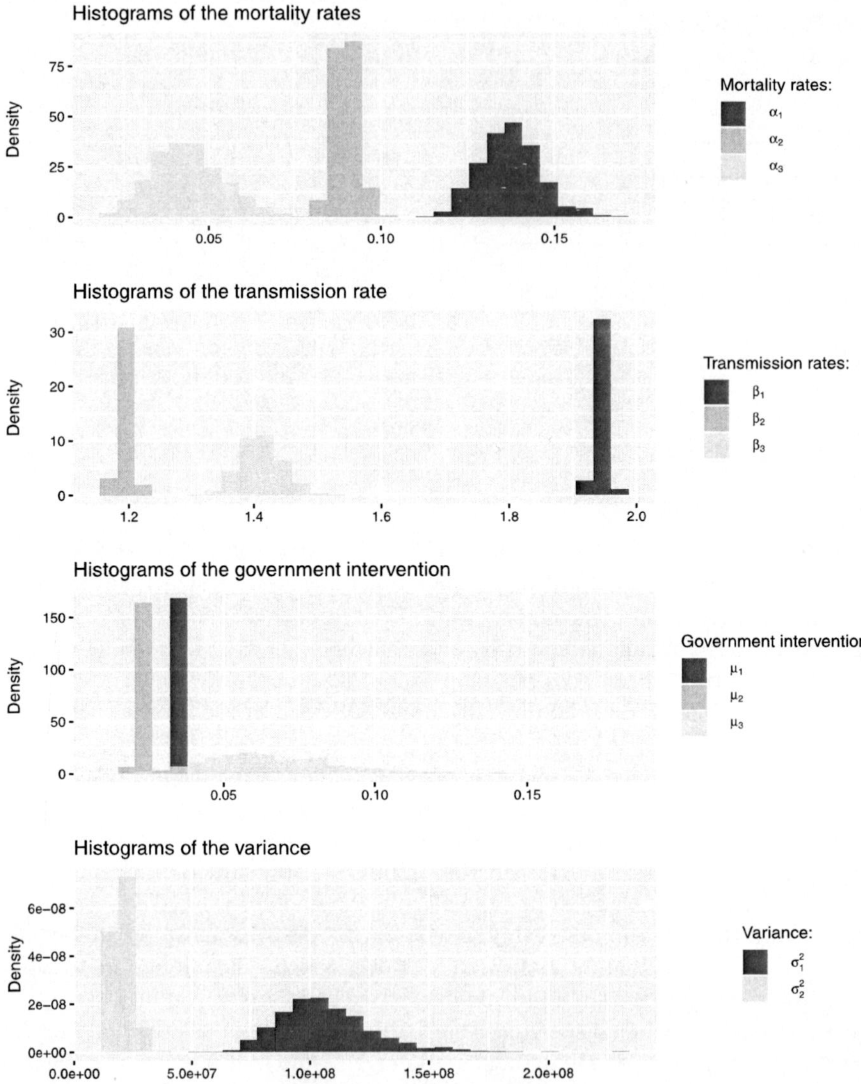

Fig. 2 This figure shows the time evolution of the posterior marginal distributions for each parameter in the dynamic model over the three waves. The first three rows show a time-shifting to the left for all parameter distributions. The fourth row shows the variance for both classes I (σ_1^2) and D (σ_2^2)

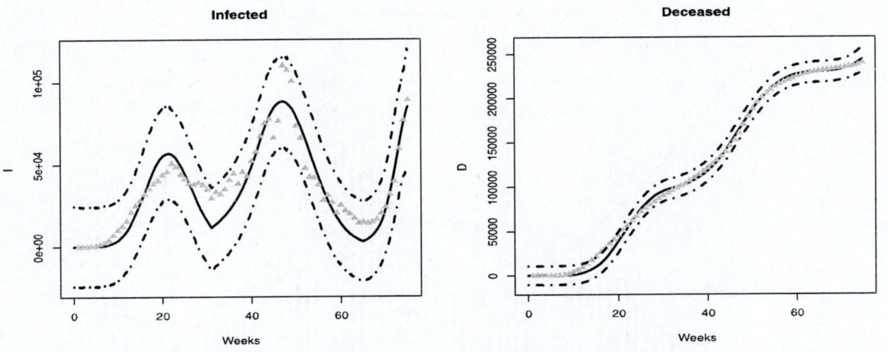

Fig. 3 The solid (black) line shows the dynamic model solution with the median estimate, the black-dotted lines envelope corresponds to the 95% HPD intervals obtained at each time point. Gray dots are data points

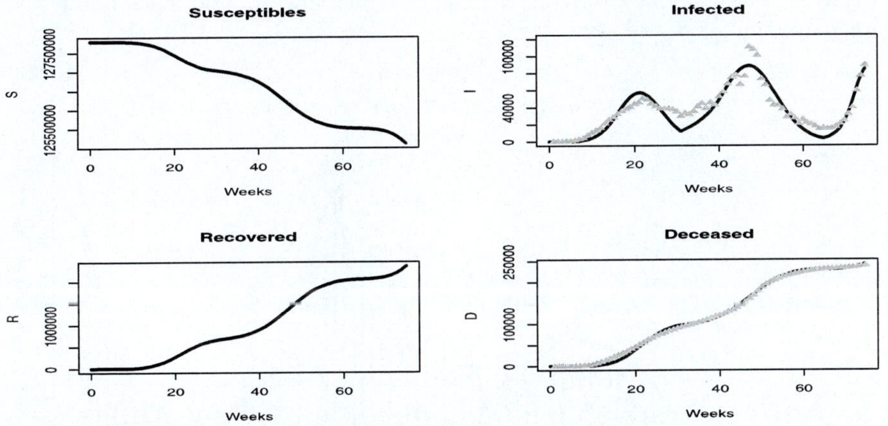

Fig. 4 This figure shows the estimates for all classes for the dynamic model, even for the non-observable classes S and R. Solid black lines are the point time estimates, and gray dots are the recorded data

One of the worthy advantages of using these dynamic models is that it allows us to estimate states that are not directly observable. In Fig. 4, we display the point inference for the susceptible and recovered classes in the dynamic epidemic model. In Fig. 5a, we display the whole evolution of the transmission rate β, representing the number of influential contacts made by each person per time. The transmission rate β distribution develops during the pandemic depicting the increasing and decreasing confirmed cases over time. We also calculate the force of infection λ displayed in Fig. 5b. We observe that the median value of the force of infection matches nearly perfectly the number of infectious.

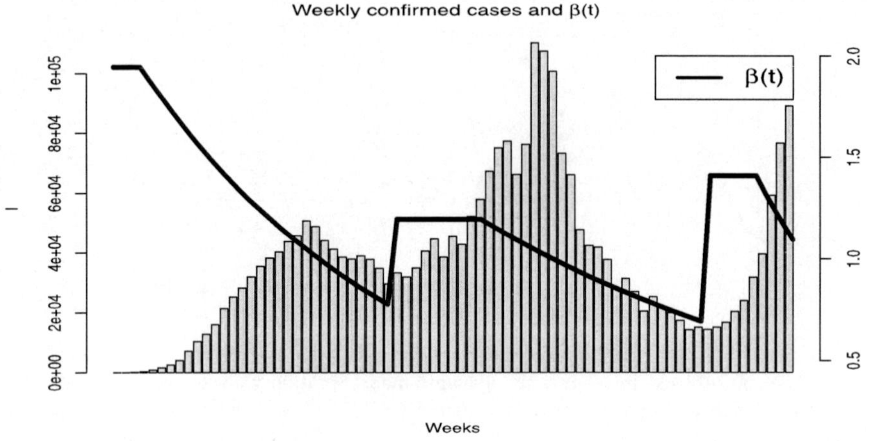

(a) Weekly data points for confirmed cases (histogram). The black line is the median value of $\beta(t)$.

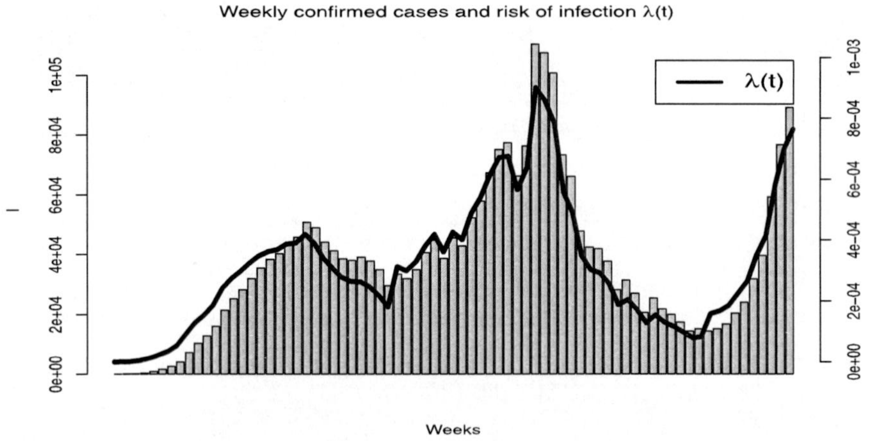

(b) Weekly data points for confirmed cases and the black line represents the median value of the force of infection $\lambda(t)$.

Fig. 5 Time history of the transmission rate, β, and the force of infection, λ, through the three epidemic waves

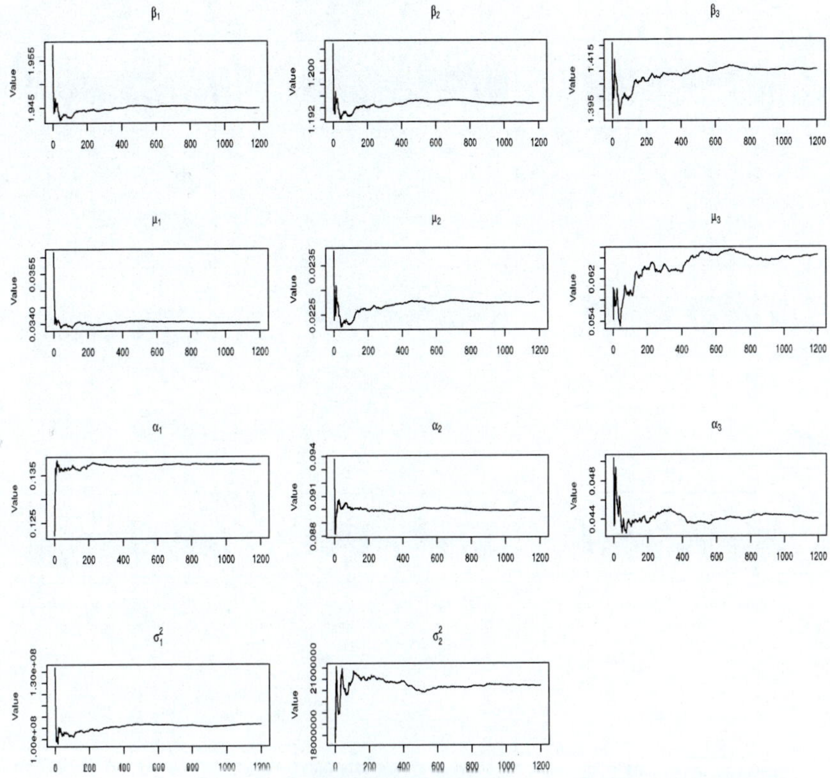

Fig. 6 The ergodic mean plots of all the parameters

Figures 6 and 7 show each model parameter's ergodic averages and chain traces. On the one hand, the traces do not show long-term trends of increase or decrease, but instead, they resemble horizontal band behaviors. On the other hand, the ergodic averages show a stable trend. These facts assure us that our MCMC procedures have converged.

Sampling points from the final distribution exhibit correlations between parameters (see Fig. 8). We can observe that parameter correlations between the transmission rate β and μ and the mortality rate α have not the same intensity over the three waves. Multiple unknown factors possibly impact how these parameters correlate, for instance, vaccination campaigns. Parameter correlations are essential estimations from an epidemiologist's viewpoint. These correlations likely relate the pandemic spreading to the government's effectiveness and efficiency of public health measures and the population response. Figure 8, third row, shows how these correlations tend to fade in the third wave.

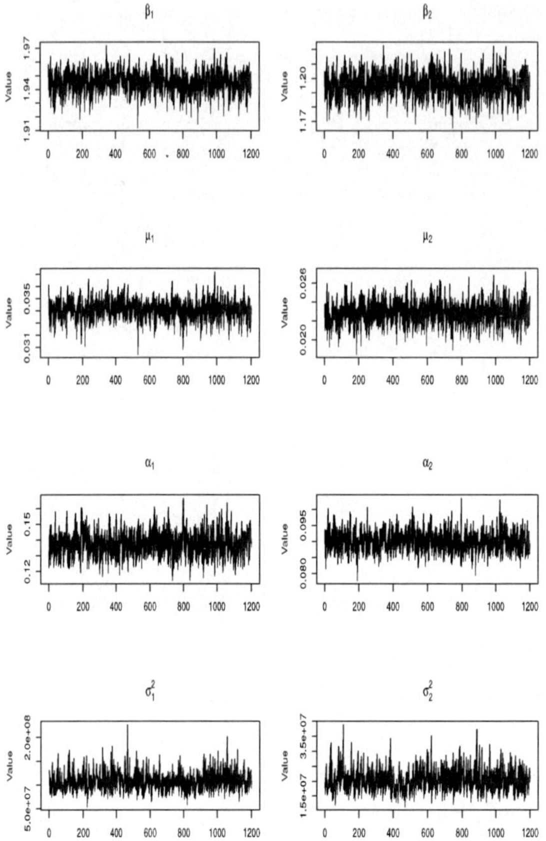

Fig. 7 Trace plots from the posterior sample of all the parameters

5 Conclusions

The central point of this work was to correlate the shifting of the mortality rate distributions during the pandemic in Mexico to the public health measures. We pointed out that multiple unknown factors possibly impacted this correlation, for instance, vaccination campaigns. In Mexico, the vaccination program started in the last week of December 2020, and mass vaccination for the older cohort population began in February 2021 when the second wave began to decline. During the first wave, misinformation dominated the public opinion respecting the pandemic. There was a lack of equipment for testing worldwide, probably resulting in a sub-registration of the number of confirmed cases. Many unknown factors impacted the parameter alpha undoubtedly. Another annotation is that the parameter correlations between mortality rates and transmission rates faded through each pandemic wave. We observe less correlation between these two parameters in the third pandemic wave. Another

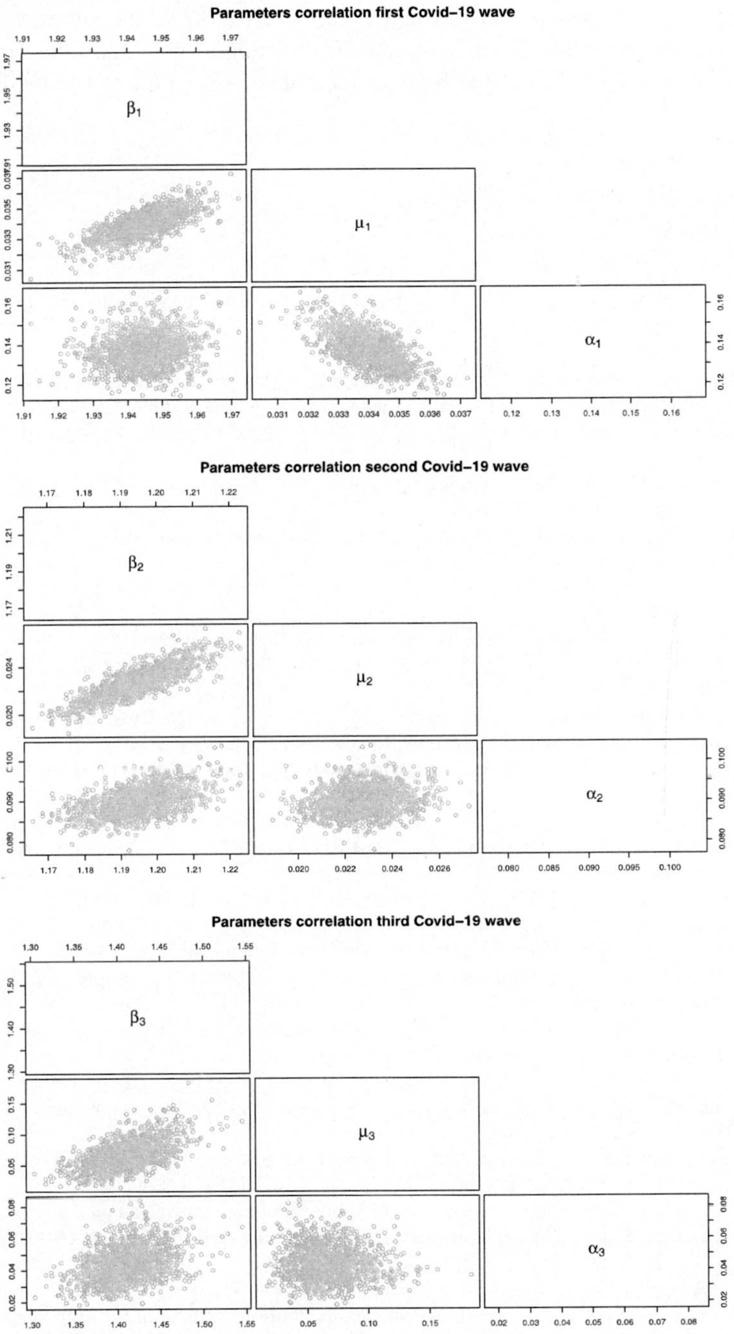

Fig. 8 Parameter correlation map through the three waves

exciting outcome from our analysis was to capture the role of the parameter β over time. The behavior of this parameter allows us to evaluate the right timing for action. We also report the change in the force of infection λ over time to see the force of infection of this new disease.

References

1. Ahmed, I., Modu, G.U., Yusuf, A., Kumam, P., Yusuf, I.: A mathematical model of Coronavirus disease (COVID-19) containing asymptomatic and symptomatic classes. Results Phys. **21**, 103776 (2021). https://doi.org/10.1016/j.rinp.2020.103776
2. Alanazi, S.A., Kamruzzaman, M.M., Alruwaili, M., Alshammari, N., Alqahtani, S.A., Karime, A.: Measuring and preventing COVID-19 using the SIR model and machine learning in smart health care. J. Healthc. Eng. 12 pp (2020). Article ID 8857346. https://doi.org/10.1155/2020/8857346
3. Anderson, R.M. (ed.): The Population Dynamics of Infectious Diseases: Theory and Applications. Springer (1982)
4. Andrieu, C., Thoms, J.: A tutorial on adaptive MCMC. Stat. Comput. **18**(4), 343–373 (2008). https://doi.org/10.1007/s11222-008-9110-y
5. Bailey, N.T.J.: The Mathematical Theory of Infectious Diseases and Its Applications, 2nd edn. Charles Griffin (1975)
6. Bjørnstad, O.N.: Epidemics: Models and Data Using R. Springer (2018)
7. Brooks, S., Gelman, A., Jones, G., Meng, X.-L.: Handbook of Markov Chain Monte Carlo. CRC Press (2011)
8. National Council for Science and Technology of Mexico (CONACyT): Covid-19 México. https://datos.covid-19.conacyt.mx/. Last Reviewed on September 2021
9. Diekmann, O., Heesterbeek, J.A.P.: Mathematical Epidemiology of Infectious Diseases. Wiley (2000)
10. Fernández-Villaverde, J., Jones, C.I.: Estimating and simulating a SIRD model of COVID-19 for many countries, states, and cities. NBER Working Papers 27128. National Bureau of Economic Research, Inc. (2020)
11. Gelman, A., Carlin, J.B., Stern, H.S., Rubin, D.B.: Bayesian Data Analysis. Chapman and Hall/CRC (2004)
12. Guerrero-Flores, E.I.: Inferencia bayesiana en modelos de ecuaciones diferenciales: Un enfoque computacional. Unpublished M.Sc. thesis. Universidad Autónoma Metropolitana-Iztapalapa, CDMX, Mexico (2021)
13. Girolami, M.: Bayesian inference for differential equations. Theor. Comput. Sci. **408**(1), 4–16 (2008). https://doi.org/10.1016/j.tcs.2008.07.005
14. Hethcote, H.W.: The mathematics of infectious diseases. SIAM Rev. **42**(4), 599–653 (2000)
15. Keeling, M.J., Rohani, P.: Modeling Infectious Diseases in Humans and Animals. Princeton (2008)
16. Lounis, M., Bagal, D.K.: Estimation of SIR models parameters of COVID-19 in Algeria. Bull. Natl. Res. Cent. **44**, 180 (2020). https://doi.org/10.1186/s42269-020-00434-5
17. Liu, Z., Magal, P., Seydi, O., Webb, G.: Predicting the cumulative number of cases for the COVID-19 epidemic in China from early data. Math. Biosci. Eng. **17**(4), 3040–3051 (2020). https://doi.org/10.3934/mbe.2020172
18. Muse, A.H., Mwalili, S., Ngesa, O., Almalki, S.J., Abd-Elmougod, G.A.: Bayesian and classical inference for the generalized log-logistic distribution with applications to survival data. Comput. Intell. Neurosci. 5820435 (2021). https://doi.org/10.1155/2021/5820435
19. Nunn, C.L., Altizer, S.: Infectious Diseases in Primates. Oxford (2006)
20. Park, S.W., Bolker, B.M.: A note on observation processes in epidemic models. Bull. Math. Biol. **82**(3), 37 (2020). https://doi.org/10.1007/s11538-020-00713-2

21. Robert, C.P.: The Bayesian Choice: From Decision-Theoretic Foundations to Computational Implementation. Springer (2001)
22. Sen, D., Sen, D.: Use of a modified SIRD model to analyze COVID-19 data. Ind. Eng. Chem. Res. (2021) https://doi.org/10.1021/acs.iecr.0c04754
23. Valderrama-Bahamndez, G.I., Fröhlich, H.: MCMC techniques for parameter estimation of ODE based models in systems biology. Front. Appl. Math. Stat. **5**, 55 (2019). https://www.frontiersin.org/article/10.3389/fams.2019.00055
24. Vera-Valdés, J.E.: The political risk factors of COVID-19. Int. Rev. Appl. Econ. **35**(2), 269–287 (2021). https://doi.org/10.1080/02692171.2020.1866973
25. Vihola, M.: On the stability and ergodicity of adaptive scaling metropolis algorithms. Stoch. Process. Appl. **121**(12), 2839–2860 (2011)
26. Vihola, M.: Robust adaptive metropolis algorithm with coerced acceptance rate. Stat. Comput. **12**(5), 997–1008 (2012). https://doi.org/10.1007/s11222-011-9269-5
27. Vynnycky, E., White, R.G.: An Introduction to Infectious Disease Modelling. Oxford University Press, Oxford (2010)
28. World Health Organization: Coronavirus Disease (COVID-19) pandemic. https://www.who.int. Last Reviewed on September 2021

Social Lag in the Municipalities of the State of Guerrero, México

María Guzmán-Martínez and Fernando Camacho-Sandoval

1 Introduction

The National Council for the Evaluation of Social Development Policy (CONEVAL) [1] creates a Social Lag Index (IRS) to order Mexico's territorial units (Entities, Municipalities, and Localities). CONEVAL points out that the IRS does not constitute a measure of poverty, since it does not include information on income, access to social security or access to food. With the help of the IRS and the Dalenius–Hodges stratification technique [2], CONEVAL classifies territorial units into five Degrees of Social Lag (GRS): Very low, Low, Medium, High, and Very high. Since 2000, CONEVAL has conducted this analysis every five years, without major changes in its methodology.

Using principal component analysis (PCA), CONEVAL creates the Social Lag Index from the first linear combination of 11 indicators: three related to educational level, one with access to health services, one quality and housing space, four with basic services in the home and two with household goods. Using PCA, the variance explained by the linear combinations decreases, in such a way that the first component explains the highest percentage of variance and the last component explains the lowest percentage of variance. In this sense, in the creation of the IRS, the total variability present in the sample is not being considered. Subsequently, CONEVAL uses the Dalenius–Hodges stratification technique [2] and the Social Lag Index created to classify all municipalities in México. Thus, the grouping of municipalities is generated taking into account only the information present in the first principal component. It should be noted that homogeneity in the municipalities of the state of Guerrero acquires a relative character when compared to the rest of the municipal-

M. Guzmán-Martínez (✉)
Universidad Autónoma de Guerrero, Chilpancingo, Mexico
e-mail: marnezmar@yahoo.com.mx

F. Camacho-Sandoval
CIMAT, Aguascalientes, Mexico

I. Antoniano-Villalobos et al. (eds.), *Interdisciplinary Statistics in Mexico*,
Springer Proceedings in Mathematics & Statistics 397,
https://doi.org/10.1007/978-3-031-12778-6_8

ities at the national level, since the geographical, social, and economic conditions are dissimilar for all the municipalities of México. In other words, more heterogeneous municipalities are grouped considering less information than is available in the sample.

According to CONEVAL [1], Guerrero has been one of the states with the highest Social Lag Index. Since 2000, Guerrero has had more than 70% of its municipalities with Medium, High and Very High IRS; this in population is equivalent to 56.77% in the year 2005, to 55.0% in the year 2010, to 59.28% in the year 2015 and 45.99% in the year 2020, that is, almost half of the population of Guerrero has limitations in education, basic services and quality of housing. In the last 15 years, the municipalities of Guerrero that are in a very high degree of social lag are: Alcozauca de Guerrero, Atlamajalcingo del Monte, Atlixtac, Copalillo, Copanatoyac, Metlatónoc, Pedro Ascencio Alquisiras, San Miguel Totolapan, Tlacoachistlahuaca, Tlacoapa, Xalpatláhuac, Xochistlahuaca, Zapotitlán Tablas, Acatepec, Cochoapa el Grande, José Joaquín de Herrera.

The objective of the work is to perform a hierarchical cluster analysis with the 11 indicators, used by CONEVAL, to classify the 81 municipalities of Guerrero based on their similarities. This can be done from the hierarchical classification generated by the hierarchical cluster analysis. This association structure is convenient since it generates a partition of the data into homogeneous groups. To determine if there is spatial dependence in each of the identified groups, the Global and Local Moran Index will be used. In this spatial analysis, we are interested to observe whether there are regions with persistent social deprivation in the period of time studied.

2 Methodology

2.1 CONEVAL

2.1.1 Principal Component Analysis

Principal component analysis (PCA) is a dimensionality reduction technique that seeks to explain the variability of the data from a reduced number of linear combinations of the given variables $X_1, \ldots, X_p$. The construction of the linear combinations is as follows; given p continuous random variables, $X_1, \ldots, X_p$ with variance–covariance matrix

$$\Sigma = \begin{pmatrix} \sigma_{11} & \sigma_{12} & \cdots & \sigma_{1p} \\ \sigma_{21} & \sigma_{22} & \cdots & \sigma_{2p} \\ \vdots & \vdots & \ddots & \vdots \\ \sigma_{p1} & \sigma_{p1} & \cdots & \sigma_{pp} \end{pmatrix}$$

where $\sigma_{ij} = Cov(X_i, X_j)$. The spectral decomposition of Σ is given by

$$\Sigma = \sum_{i=0}^{p} \lambda_i \mathbf{e}_i \mathbf{e}_i' = \lambda_1 \mathbf{e}_1 \mathbf{e}_1' + \lambda_2 \mathbf{e}_2 \mathbf{e}_2' + \cdots + \lambda_p \mathbf{e}_p \mathbf{e}_p'$$

where $\mathbf{e}_i' = (e_{i1}, \ldots, e_{ip})$. Based on the eigenvalues and eigenvectors of Σ, the p linear combinations (principal component, CP), $Y_1, \ldots, Y_p$, are constructed as follows:

$$\begin{aligned} Y_1 &= e_{11}X_1 + e_{12}X_2 + \cdots + e_{1p}X_p \\ Y_2 &= e_{21}X_1 + e_{22}X_2 + \cdots + e_{2p}X_p \\ &\vdots \\ Y_p &= e_{p1}X_1 + e_{p2}X_2 + \cdots + e_{pp}X_p \end{aligned}$$

where $Var(Y_i) = \mathbf{e}_i' \Sigma \mathbf{e}_i = \lambda_i,\ i = 1, \ldots, p$. It is worth mentioning that the p linear combinations are incorrectly related. The PCA assures that

$$Var(Y_1) \geq Var(Y_2) \geq \cdots \geq Var(Y_p) \geq 0$$

That is, $\lambda_1 \geq \lambda_2 \geq \cdots \geq \lambda_p \geq 0$. Note that

$$\sum_{i=1}^{p} Var(Y_i) = \lambda_1 + \cdots + \lambda_p = \sum_{i=1}^{p} Var(X_i) = \sigma_{11} + \cdots + \sigma_{pp}.$$

Thus the variance explained by the k-th CP, $k = 1, \ldots, p$, is given by

$$\frac{\lambda_k}{\sum_{i=1}^{p} \lambda_i}$$

This means that each linear combination constructed, Y_i, explains only a total percentage of the total variance of the sample, being the first linear combination Y_1, the one that explains the highest percentage of variance of the sample.

2.1.2 Dalenius–Hodges Method

The Dalenius–Hodges (1959) method is a univariate stratification technique that helps with the formation of homogeneous groups, such that the variance within groups is minimal and between groups is maximal. The range $x_L - x_0$ of the variable to be stratified, X, is divided into L parts; at points $x_1 < \cdots < x_h < x_{L-1}$; each part corresponds to a stratum, i.e., the given variable X is used as a stratification variable. In the Dalenius–Hodges method, also known as the cum$\sqrt{(f)}$ rule or the cumulative root frequency method. The method is as follows [3]:

- Arrange the stratification variable X in ascending order;

- Group the X into a number of classes, J;
- Determine the frequency in each class $f_i, i = 1, 2, \ldots, J$;
- Determine the square root of the frequencies in each class;
- Cumulative the square root of the frequencies

$$\sum_{i=1}^{J} \sqrt{f_i}$$

- Divide the sum of the square root of the frequencies by the number of strata

$$Q = \frac{1}{L} \sum_{i=1}^{J} \sqrt{f_i}$$

- Take the upper boundaries of each stratum to be the x values corresponding to

$$Q, 2Q, 3Q, \ldots, (L-1)Q, LQ$$

According to Er [4], the main problem with the Dalenius–Hodges method is the arbitrariness in deciding the number of classes, L; she also mentions that the method is highly dependent on the uniform distribution within the strata. Hedlin [5] said that the final stratum boundaries depend on the initial choice of the number of classes and there is no theory that gives the best number of classes.

For the construction of the strata, CONEVAL first generates the p linear combinations, $Y_1, Y_2, \ldots, Y_p$; then with the first linear combination, Y_1, construct the IRS as follows

$$IRS = (Y_1 - mean(Y_1))/\sqrt{Var(Y_1)}$$

It then uses the Dalenius–Hodges method to stratify the IRS variable, resulting in the GRS with five strata: Very low, Low, Medium, High, and Very higt.

2.2 *Clustering*

2.2.1 Agglomerative Hierarchical Method

Cluster analysis aims to group elements into homogeneous groups based on their similarities. Cluster analysis studies three types of problems: data partitioning, variable classification, and hierarchical construction. In the construction of hierarchies, the aim is to build clusters of observations in a hierarchical way by similarity. In a hierarchical classification, the data is arranged in levels, so that the higher levels contain the lower levels. This construction method does not define groups, rather it explains a chain association structure that can exist between the data. However, the hierarchy built allows a partition of the observations to be obtained, giving rise to a natural grouping of the data. This cluster analysis has already been used to stratify the population according to their socioeconomic situation, at the Basic Geostatistical Area (AGEB)[1] and urban block scale for México City [6] and for the city of Aguascalientes [7].

Clustering techniques can be classified into hierarchical division methods and agglomerative hierarchical methods. In hierarchical division methods, there is only one group of objects, which is divided into two groups with the restriction that the two new groups are separated as much as possible; the process continues until there are as many groups as there are objects in the sample. Agglomerative algorithms start with many clusters (one singular cluster for each element of the sample). Similar clusters are grouped on each iteration until all observations form a unique set.

Hierarchical clustering method:

1. Select the p variables that will provide the information of the N objects in the sample.
2. Choose the measure of similarity, which must take into account the nature of the variables, their scale, and the nature of the data. Generally, when objects are grouped, the measure of similarity is a distance function for the p-dimensional observations. Metrics include Euclidean, Minkowski, and Mahalanobis distance [8]. Before calculating the distance matrix, it is recommended to standardize the variables to remove the effect of the scale.
3. Select linkage methods examples: single linkage, complete linkage, average linkage. The more popular is Ward which considers hierarchical clustering procedures based on minimizing the loss of information from joining two groups. This method is usually implemented with a loss of information taken to be an increase in an Error Sum of Squared criterion [8].
4. Identify the clusters, for this, we can use the average values of the p variables in each of the identified groups.
5. Finally, it is validated that the groupings have a meaning that verifies or not the initial hypotheses and is congruent with the criteria of homogeneity between and within the groups.

The Euclidean distance and the Ward method are used for the study. The Euclidean distance between two p-dimensional observations $\mathbf{x}' = (x_1, \ldots, x_p)$ and $\mathbf{y}' = (y_1, \ldots, y_p)$ is given by

$$d(\mathbf{x}, \mathbf{y}) = \sqrt{(x_1 - y_1)^2 + (x_2 - y_2)^2 + \cdots + (x_p - y_p)^2}$$

[1] According to the definition of the National Institute of Statistics and Geography (INEGI), an urban AGEB is a geographical area occupied by a set of blocks perfectly delimited by streets, avenues, walkways or any other easily identifiable feature on the ground and whose land use is mainly residential, industrial, services, commercial, etc., and they are only assigned to the interior of urban areas, which are those with a population greater than or equal to 2,500 inhabitants and in the municipal capitals.

Ward's method initially each cluster consists of a single object, for N objects, $ESS_k = 0, k = 1, \ldots, N$, where ESS_k be the sum of the squared deviations of every object in the cluster from the cluster mean (centroid). If there are currently K clusters, define $ESS = ESS_1 + \cdots + ESS_k$. When all the cluster are combined in a single group of N object, the value of ESS is given by

$$ESS = \sum_{j=1}^{N} (\mathbf{x}_j - \bar{\mathbf{x}})'(\mathbf{x}_j - \bar{\mathbf{x}})$$

Where $\mathbf{x}_j$ is the multivariate measurement associated with the j-th object and $\bar{\mathbf{x}}$ is the mean of all the objects.

For the hierarchical clustering analysis, the *hclust* function of the *Stats* package was used [9]. To determine the number of groups, the *NbClust* function of the *NbClust* package was used [10]. Both packages can be found in the statistical software R [11].

2.2.2 Measures of Spatial Association

Moran's I was obtained as an indicator of global spatial autocorrelation [12], this index is given by

$$I = \frac{n \sum_i^n \sum_j^n W_{ij}(Y_i - \bar{Y})(Y_j - \bar{Y})}{\sum_i^n \sum_j^n W_{ij} \sum_i^n (Y_i - \bar{Y})^2} \tag{1}$$

where:

- I is restricted to the interval $[-1, 1]$.
- Y_i: is the value of the variable in each spatial unit analyzed i, in this case, the municipalities.
- Y_j: is the value of the variable in each neighboring spatial unit j.
- $\bar{Y}$: is the average of the Y.
- W_{ij}: is the matrix of weights between the spatial units i and j. When location i and j are neighbors then $W_{ij} = 1$ and $W_{ij} = 0$ when they are not, for $i = j$ we take $W_{ij} = 0$. The weight, W_{ij}, between district i and j are row-standardized (i.e., they sum to 1). In this analysis, coefficient I was computed using adjacency (queen's case contiguity, that is, the district share edges and vertices).
- n: is the number of spatial units into which the study region is divided.

When the values of Moran's I tend to -1, it is said that there is the dispersion of the data when they tend to 1, it is said that there is a perfect correlation, a value close to zero of Moran's I indicates that the distribution of the population under study is random and has no relation to the spatial pattern of the rest of the population.

Local Indicators of Spatial Autocorrelation (LISA) is used to identify local clusters and local spatial outliers [13, 14]. This statistic is given by

$$I_i = \frac{(Y_i - \bar{Y}_i)}{\sigma} \sum_j W_{ij} \frac{(Y_j - \bar{Y}_j)}{\sigma} \tag{2}$$

where σ is the standard deviation of the variable Y which, in this case, represents each of the five strata, Very low, Low, Medium, High, and Very high.

LISA is a decomposition of Moran's I used to identify the contribution of each municipality in the statistic [15]. LISA was used to derive significant spatial clustering through four cluster types: High-High, Low-Low, High-Low, and Low-High. LISA examines municipalities in relation to the characteristics of surrounding municipalities and highlights the presence of spatial clustering.

The hypothesis set for both statistics is given by:
H_0: The spatial pattern is random; H_1: The spatial pattern is not random.
H_0 is rejected if p-valor $< \alpha$, where $\alpha = 0.05$.
Since all spatial analyses require spatial weights, we obtained queen contiguity weights [15], in order to find any point of contact between the municipalities. The calculation of Moran's I and Lisa was carried out with GeoDa version 1.18.0. [16].

3 Application

The state of Guerrero is divided into seven regions; Acapulco, Costa Chica, Costa Grande, Centro, La Montaña, Norte and Tierra Caliente (Fig. 1). The 81 municipalities of the state of Guerrero are divided into seven regions.

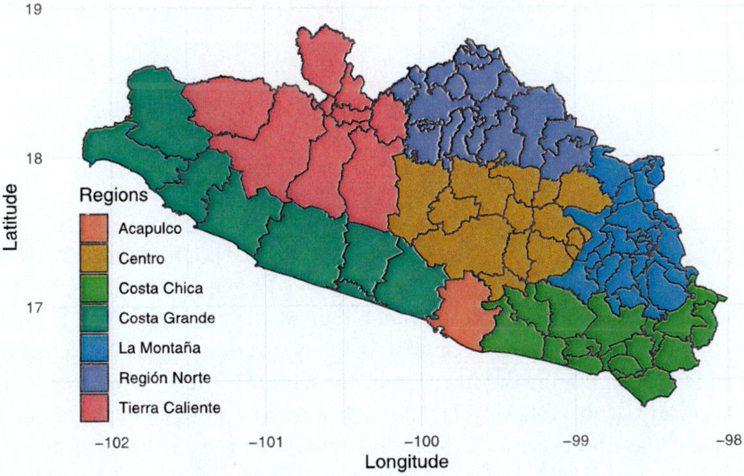

Fig. 1 Regions and municipalities of the state of Guerrero

The municipalities and the population of the regions are shown in Table 1. According to the 2020 Population and Housing Census of the National Institute of Statistics and Geography [17], Guerrero had 3,540,686 inhabitants; the Acapulco region has the largest population (779,566 inhabitants), while Tierra Caliente is the least populated region (244,555).

For the analysis of hierarchical clusters, the 11 indicators used by CONEVAL are used:
Educational deprivation:

- Illiterate population 15 years old and over (X_1).
- Population 6 – 14 years old not attending school (X_2).
- Population 15 years of age and older with incomplete basic education (X_3).

Access to health services:

- Population with no access to health services (X_4).

Quality and spaces of the house:

- Homes with dirt floors (X_5).

Basic services in homes:

- Dwellings without toilet or sanitary facilities (X_6).
- Homes that do not have piped water from the public network (X_7).
- Homes that do not have drainage (X_8).
- Homes that do not have electricity (X_9).

Household goods:

- Homes that do not have a washing machine (X_{10}).
- Homes that do not have a refrigerator (X_{11}).

4 Results

4.1 CONEVAL

Tables 2 and 3 show the number of municipalities and total population that CONEVAL identified with Very low, Low, Medium, High, and Very high social lag. In general, the number of municipalities and total inhabitants with very high social lag has been maintained in the last 15 years. In contrast, in the High stratum, the number of municipalities and total population did not remain constant; by 2020, only 8 municipalities were in this group, with a total of 321,190 people with high social deprivation, almost a third of what was observed in 2015. This would mean good news for Guerrero. For the years, 2015 and 2020, the CONEVAL grouped Guerrero's municipalities into

Table 1 Population and municipalities of the regions of the state of Guerrero

Regions	Population in 2020	Municipalities (Municipality Code)
Acapulco	779,566	Acapulco (001)
Costa Chica	461,500	Ayutla de los Libres (012), Azoyú (013), Copala (018), Cuajinicuilapa (023), Cuautepec (025), Florencio Villarreal (030), Igualapa (036), Juchitán (080), Marquelia (077), Ometepec (046), San Luis Acatlán (052), San Marcos (053), Tecoanapa (056), Tlacoachistlahuaca (062), Xochistlahuaca (071)
Costa Grande	423,756	Atoyac de Álvarez (011), Benito Juárez (014), Coahuayutla de José María Izazaga (016), Coyuca de Benítez (021), Petatlán (048), Tecpan de Galeana (057), La Unión de Isidoro Montes de Oca (068), Zihuatanejo de Azueta (038)
Centro	725,798	Ahuacuotzingo (002), Chilapa de Álvarez (028), Chilpancingo de los Bravo (029), Eduardo Neri (075), Gral. Heliodoro Castillo (032), José Joaquín de Herrera (079), Juan R. Escudero (039), Leonardo Bravo (040), Mártir de Cuilapan (042), Mochitlán (044), Quechultenango (051), Tixtla de Guerrero (061), Zitlala (074)
La Montaña	406,048	Acatepec (076), Alcozauca de Guerrero (004), Alpoyeca (005), Atlamajalcingo del Monte (009), Atlixtac (010), Cochoapa el Grande (078), Copanatoyac (020), Cualac (024), Huamuxtitlán (033), Iliatenco (081), Malinaltepec (041), Metlatónoc (043), Olinalá (045), Tlacoapa (063), Tlalixtaquilla de Maldonado (065), Tlapa de Comonfort (066), Xalpatláhuac (069), Xochihuehuetlán (070), Zapotitlán Tablas (072)
Norte	499,462	Apaxtla de Castrejón (006), Atenango del Río (008), Buenavista de Cuéllar (015), Cocula (017), Copalillo (019), Cuetzala del Progreso (026), General Canuto A. Neri (031), Huitzuco de los Figueroa (034), Iguala de la Independencia (035), Ixcateopan de Cuauhtémoc (037), Pedro Ascencio de Alquisiras (047), Pilcaya (049), Taxco de Alarcón (055), Teloloapan (058), Tepecuacuilco de Trujano (059), Tetipac (060)
Tierra Caliente	244,555	Ajuchitlán del Progreso (003), Arcelia (007), Coyuca de Catalán (022), Cutzamala de Pinzón (027), Pungarabato (050), San Miguel Totolapan (054), Tlalchapa (064), Tlapehuala (067), Zirándaro (073)

Table 2 CONEVAL: municipalities with GRS, 2005-2020

GRS	Municipalities			
	2005	2010	2015	2020
Very low	2	4	0	0
Low	8	7	8	21
Medium	24	26	24	31
High	25	26	32	8
Very high	22	18	17	21
Total	81	81	81	81

Source CONEVAL [1]

Table 3 CONEVAL: population with GRS, 2005 -2020

GRS	Municipalities			
	2005	2010	2015	2020
Very low	233,053	1,290,262	0	0
Low	1,113,629	234,804	1,438,730	1,912,444
Medium	675,603	715,830	776,928	869,223
High	705,718	817,590	996,688	321,190
Very high	387,199	330,282	320,905	437,828
Total	3,115,202	3,388,768	3,533,251	3,540,685

Source CONEVAL [1]

four groups: Low, Medium, High, and Very high. In Fig. 2, it can be observed that most of the municipalities with very high social lag are located in the region of La Montaña; from 2005 to 2015, the municipalities with medium and high social lag are scattered in the regions of Costa Grande, Centro, Norte, and Tierra Caliente. Only in 2020, a more structured grouping of municipalities is observed. In this year, most of the municipalities with a low degree of social deprivation are located in the Costa Grande and Centro regions.

4.2 Cluster

The following are the results of the hierarchical cluster analysis using the Ward method and the Euclidean distance for the years 2005, 2010, 2015 and 2020. The grouping of the municipalities is obtained from the partition of the data generated from the hierarchical classification of the data by the similarity of the hierarchical cluster analysis.

For 2005 and 2010, the municipalities were grouped into Very low, Low, Medium, High, and Very high degree of social lag. Tables 4 and 5, respectively, show the groups

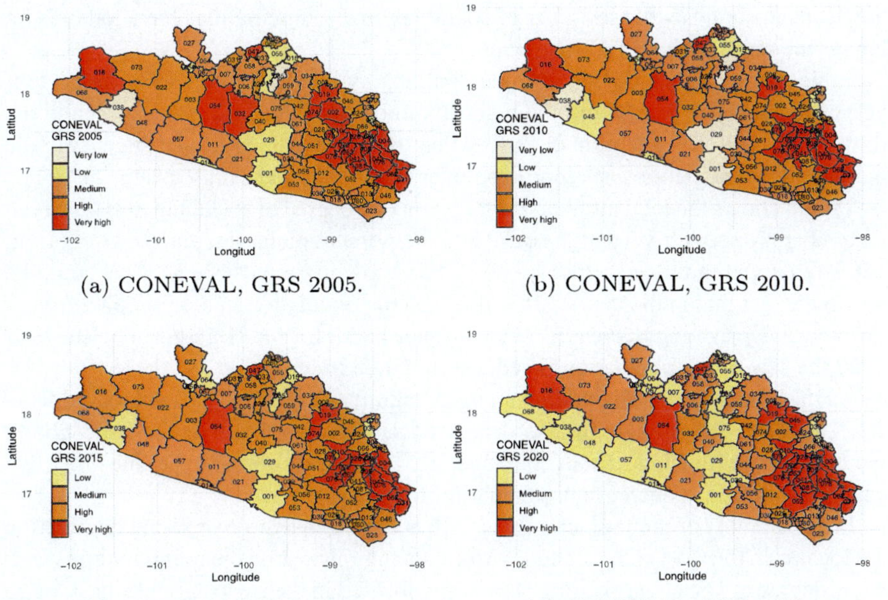

(a) CONEVAL, GRS 2005. (b) CONEVAL, GRS 2010.

(c) CONEVAL, GRS 2015. (d) CONEVAL, GRS 2020.

Fig. 2 Coneval, 2005–2010

Table 4 Averages of the 11 indicators in the groups of municipalities in the state of Guerrero, 2005

GRS	X_1	X_2	X_3	X_4	X_5	X_6	X_7	X_8	X_9	X_{10}	X_{11}
Very low	9.4	4.7	41.8	56.3	15.5	11.6	21.0	7.8	4.9	50.4	17.3
Low	21.6	6.6	65.0	84.3	29.5	26.5	32.9	29.5	5.1	67.5	30.8
Medium	32.4	9.9	79.2	94.6	53.7	48.6	63.7	59.5	11.5	84.4	52.4
High	27.9	4.8	72.2	80.4	68.0	60.6	46.6	73.6	32.4	96.2	73.9
Very high	54.2	15.8	88.3	96.3	81.3	74.3	36.5	82.2	23.9	98.3	84.0

created and the averages of the 11 indicators for each of the groups. The groups with the highest degree of social lag have the highest averages in the indicators.

For the years 2015 and 2020, it was decided to group the municipalities in low, medium, high and very high degree of social deprivation, in order to compare the structure (pattern) of grouping the municipalities of the hierarchical cluster analysis with that generated by the CONEVAL. Then for 2015 and 2020, the municipalities were grouped into Low, Medium, High, and Very high GRS. Tables 6 and 7, respectively, show the average values of the 11 indicators for the four groups created. The highest averages of 11 indicators are found in the groups with the greatest social lag.

Comparing the averages of the 11 indicators in each of the groups obtained in the four years of the study, these averages are observed to have been decreasing over time. That is, the level of the degree of social lag has decreased in the last 15 years, this leads to suppose that the level of social lag that municipalities have today is no longer the same as that of 15 years ago.

Table 8, shows the number of municipalities by degree of social lag in the four years of study. The group with the highest number of municipalities is typically the Low group with the exception of 2010 when it is the Medium group. In general, the number of municipalities has increased over time in the Very high group.

Table 9 shows the total number of inhabitants by degree of social lag. It is observed that in the group with Very high social lag, the population has increased, going from 138,947 people in the year 2005 to 677,712 people in the year 2020. This could be due to the fact that in the years 2015 and 2020 the population in the group with Very low social deprivation decreased. On the other hand, for the High group in the year 2020 the size of the population is reduced to 64,282 population.

As the number of municipalities in the Medium and High groups decreased from 2015 to 2020, the population also decreased in those groups. For the High group, the number of municipalities in 2010 and 2020 decreased compared to the other two years, resulting in a decrease in population for those years.

In Fig. 3, most of the municipalities with a very high degree of social deprivation are located in the Costa Chica and La Montaña regions; while the municipalities with a low degree of social deprivation are mainly located in the Costa Grande and Centro regions. It is also observed that in general, the municipalities with a medium degree of social deprivation are located in the Tierra Caliente region. The grouping of the 81 municipalities with the hierarchical cluster analysis is much better structured than the one observed in the CONEVAL results.

Table 5 Averages of the 11 indicators in the five groups of municipalities in the state of Guerrero, 2010

GRS	X_1	X_2	X_3	X_4	X_5	X_6	X_7	X_8	X_9	X_{10}	X_{11}
Very low	13.4	5.5	52.8	42.8	15.5	12.2	29.3	12.4	2.1	48.8	18.0
Low	17.4	6.1	58.7	52.5	26.7	17.4	39.3	19.6	2.6	67.0	25.4
Medium	22.6	6.3	66.6	48.2	15.3	29.1	58.8	34.1	5.3	67.9	30.6
High	23.2	4.7	67.9	50.3	34.0	46.0	57.6	59.9	25.8	92.0	67.5
Very high	39.5	11.7	81.9	51.9	28.2	47.9	51.5	59.9	10.3	89.8	61.9

Table 6 Averages of the 11 indicators for the groups of municipalities in the state of Guerrero, 2015

GRS	X_1	X_2	X_3	X_4	X_5	X_6	X_7	X_8	X_9	X_{10}	X_{11}
Low	12.2	3.3	48.9	14.4	9.2	9.3	11.4	10.5	1.4	49.4	15.8
Medium	17.5	3.8	58.9	13.0	13.8	16.9	25.0	21.5	2.2	67.2	25.0
High	26.0	7.7	72.4	15.2	22.7	25.6	33.4	34.7	4.2	70.0	35.3
Very high	30.4	5.0	69.9	11.3	31.9	35.4	14.4	51.5	9.9	92.3	62.5

Table 7 Averages of the 11 indicators for the groups of municipalities in the state of Guerrero, 2020

GRS	X_1	X_2	X_3	X_4	X_5	X_6	X_7	X_8	X_9	X_{10}	X_{11}
Low	13.1	5.2	46.6	23.1	11.9	8.6	11.7	10.6	1.3	53.5	17.6
Medium	14.4	8.9	53.2	37.4	11.9	9.4	9.3	9.3	1.7	43.4	12.6
High	24.8	9.1	66.7	20.6	19.7	16.6	57.4	26.2	2.2	75.0	28.7
Very high	26.1	8.9	63.4	15.8	33.0	25.5	12.8	38.4	5.3	84.1	48.2

Table 8 Cluster: Municipalities with GRS, 2005–2020

GRS	Municipalities			
	2005	2010	2015	2020
Very low	4	17	0	0
Low	36	13	27	42
Medium	23	25	24	11
High	9	7	12	4
Very high	9	19	18	24
Total	81	81	81	81

Table 9 Cluster: Population with GRS, 2005-2020

GRS	Population			
	2005	2010	2015	2020
Very low	1,165,038	1,694,588	0	0
Low	1,027,516	434,209	2,045,795	2,532,349
Medium	524,448	629,392	688,713	266,342
High	259,253	189,835	346,609	64,282
Very high	138,947	440,744	452,134	677,712
Total	3,115,202	3,388,768	3,533,251	3,540,685

In Fig. 3, the growth of the Very high group can be graphically observed, as shown in Table 8. In Fig. 3, it can be observed that the 9 municipalities with very high social deprivation in 2005 were maintained until 2020; it can also be observed the addition of more municipalities to the Very high group in the following years.

4.2.1 Spatial Correlation Analysis

Table 10 shows the calculation of the Moran Index by year for the five strata identified in the hierarchical cluster analysis. At the significance level of $\alpha = 0.05$; 15 of the 20 correlations calculated are statistically significant. This means that the grouped municipalities present a spatial dependence.

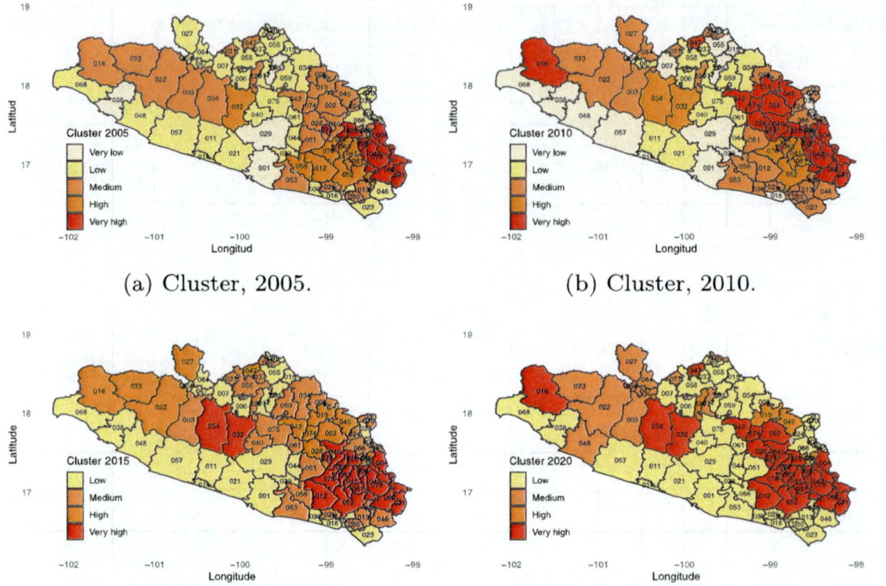

(a) Cluster, 2005.

(b) Cluster, 2010.

(c) Cluster, 2015.

(d) Cluster, 2020.

Fig. 3 Cluster, 2005-2020

Table 10 Moran's I, 2005-2020

Cluster	Moran's I			
	2005	2010	2015	2020
Very low	0.072*	0.083*	0	0
Low	0.006	0.130*	0.129*	0.147*
Medium	0.115*	0.149*	−0.038	0.508*
High	0.250*	0.182*	0.111*	0.087
Very high	0.392*	0.237*	0.275*	0.224*

*Statistically significant at 0.05

The calculation of LISA for the group with Very high social lag, was statistically significant for the four years of the study; 2005 (LISA = 0.392, $p\text{-}value = 0.001$), 2010 (LISA = 0.237, $p\text{-}value = 0.001$), 2015 (LISA = 0.275, $p\text{-}value = 0.002$) and 2020 (LISA = 0.224, $p\text{-}value = 0.003$). Figures 4 and 5 show the grouping of municipalities with very high social lag into four groups High–High, Low–Low, High–Low, and Low–High. In this classification, two groups are important High–High and Low–High; the High–High group indicates that municipalities with high social lag tend to be surrounded by municipalities with high social lag; the Low–High group indicates that the municipalities with low social lag are surrounded by municipalities with high social lag. The Low–High group is transitional, this is, the municipalities in this group will at a certain point change to the High–High group, which is the group

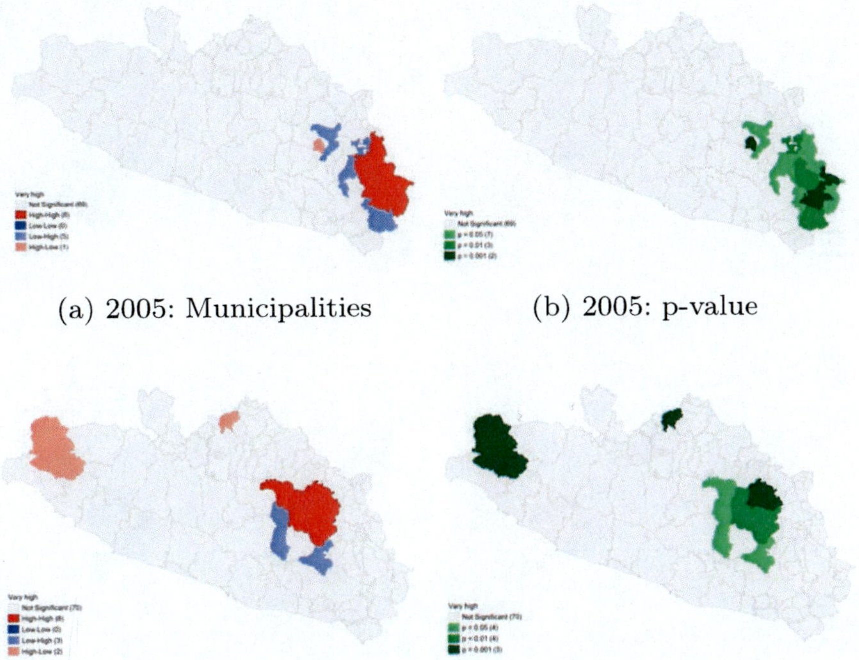

(a) 2005: Municipalities (b) 2005: p-value

(c) 2010: Municipalities (d) 2010: p-value

Fig. 4 Municipalities with very high social deprivation (2005 and 2010)

where the municipalities with the highest levels of social deprivation are located. In Figs. 4 and 5, the *p*-values of the High–High, Low–Low, High–Low, and Low–High groupings are also shown, all statistically results significant with $\alpha = 0.05$.

4.3 Comparison

According to the results of the hierarchical cluster analysis (Table 9), in 2015 and 2020, the population of the Low group increased more than reported by CONEVAL (Table 3). On the other hand, with the hierarchical cluster analysis, the population of the Medium group decreased from 2015 to 2020, while according to CONEVAL there was an increase in population in that group (Tables 3 and 9). For the years 2010 and 2020, the results of the hierarchical cluster analysis show that the population of the High group was considerably reduced compared to that reported by CONEVAL for the same group. The above indicates that there are differences in the grouping structures of the municipalities, and therefore the total population in the strata changes.

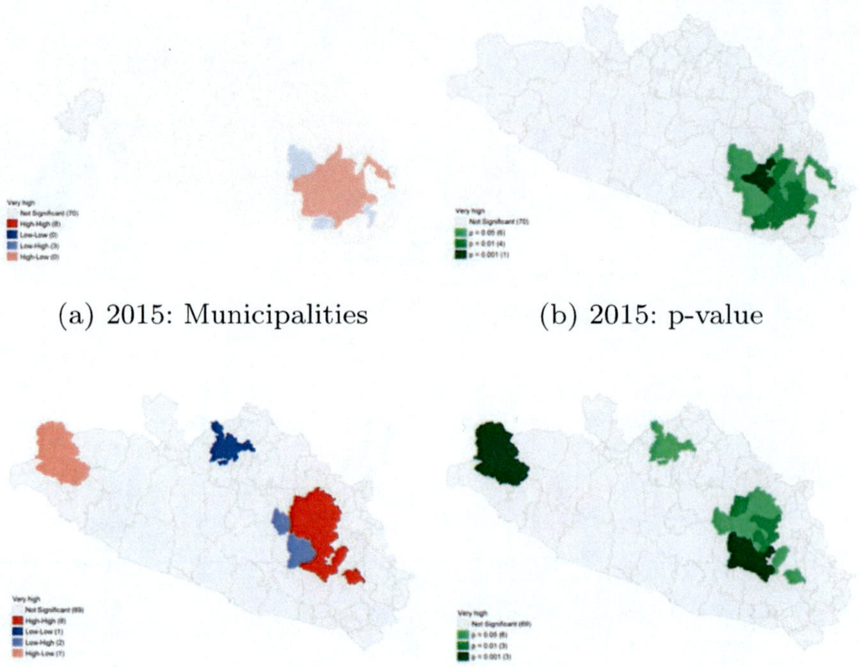

(a) 2015: Municipalities (b) 2015: p-value

(c) 2020: Municipalities (d) 2020: p-value

Fig. 5 Municipalities with very high social deprivation (2015 and 2020)

If we identify the coincidences of both methodologies in the Very high group, which is the group of greatest interest because it contains the municipalities with the highest degree of social deprivation, we have the following results. The 9 municipalities identified by the hierarchical cluster analysis in the year 2005 were also identified by CONEVAL; for the year 2010 both methods coincide in 14 municipalities, in 2015 they coincide in 13 municipalities and in the year 2020 they coincide in 18 municipalities. That is to say, with the exception of 2005, the results of the hierarchical cluster analysis and CONEVAL are very close for the Very high group.

Figure 6 shows that the number of municipalities and the percentage of the population with very high social lag has increased in the last 15 years.

Figure 7 shows the number of inhabitants and the percentage of inhabitants for the High and Very high strata. For the year 2020, the results of the hierarchical cluster analysis coincide with the results of CONEVAL. On the other hand, for the years 2005, 2010 and 2015, the hierarchical cluster analysis assigns a lower number of inhabitants to the High and Very high strata. This may be due to the fact that the cluster analysis is only working with the information of the 81 municipalities in the state of Guerrero, instead CONEVAL is considering all municipalities in México.

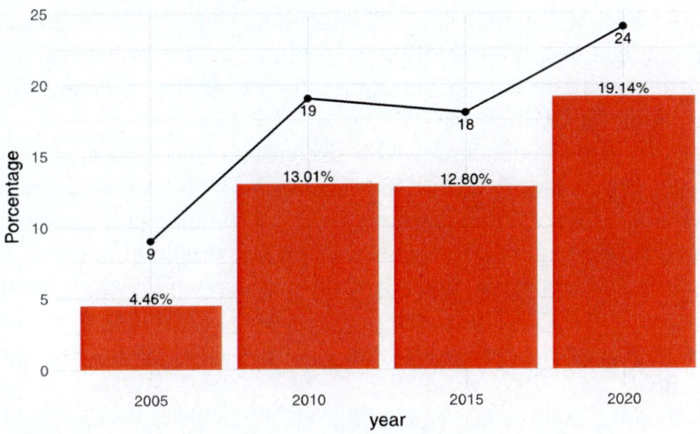

Fig. 6 Municipalities and percentages of the population with social lag very high, 2005–2020

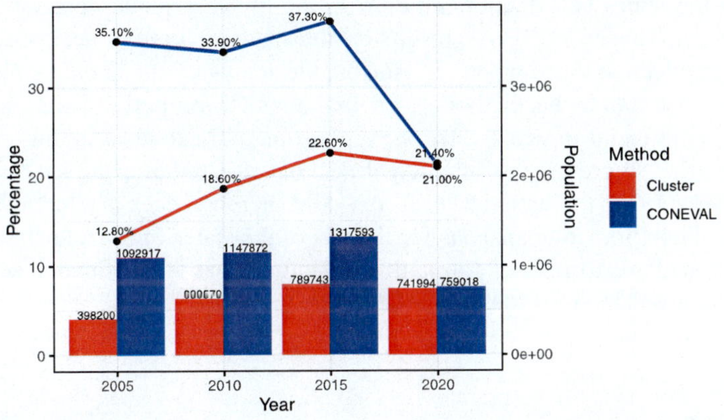

Fig. 7 Number of inhabitants and percentages of population with social lag high and very high, 2005–2020

5 Conclusions

The results obtained with the hierarchical cluster analysis for the state of Guerrero, keep a certain relation with those presented with the CONEVAL at national level, particularly with the 2020 results.

The grouping of the municipalities of Guerrero with the analysis of hierarchical clusters can be an alternative to the methodology used by CONEVAL, mainly for two reasons: the grouping of the data, obtained from the hierarchy constructed by the method, is carried out taking into account for the natural grouping that exists in the sample; and for the construction of the groups, the total variability present in the sample is taken into account.

The identification of a statistical method that helps with the correct characterization of the municipalities in terms of the degree of social deprivation they present, can be the difference between concentrating efforts to achieve a positive change, or working hard but with little results.

The study of spatial dependence, with the groups found with the hierarchical cluster analysis, shows that there is a spatial association in the municipalities. This means that the degree of social deprivation that a given municipality presents will tend to be presented by its neighbor, thus creating a vicious cycle of poverty. Therefore, it is no coincidence that the nine municipalities with very high social deprivation in 2005 will remain in this condition until the year 2020. According to the results, the greatest social lag is presented by the municipalities of La Montaña region, Costa Chica and Centro.

Although it is true that the averages of the 11 indicators have a decreasing trend, which may be due to the improvement conditions of the last fifteen years for the different municipalities of Guerrero; CONEVAL indicates that the state of Guerrero is among the states and municipalities with the greatest social deprivation at the national level. These CONEVAL results for the state of Guerrero may be due to the weighting effect that is had when carrying out the analysis with all the municipalities of the country and/or the methodology implemented, properly in the analysis of principal components in which only one proportion of the total variability present in the sample for the grouping of municipalities. In order to know the situation of the state of Guerrero at the national level, it would then be necessary to carry out the hierarchical conglomerate analysis for the rest of the states of Mexico.

Cluster analysis can be an alternative to study social lag at smaller geographic scales from municipalities to urban blocks.

References

1. The National Council for the Evaluation of Social Development Policy (CONEVAL). https://www.coneval.org.mx/Medicion/IRS/Paginas/Indice_Rezago_Social_2020.aspx
2. Dalenius, T., Hodges, Jr, J.L.: Minimum variance stratification. J. Am. Stat. Assoc. **54**(285), 88–101 (1959)
3. Gunning, P., Horgan, J.M., Yancey, W.: Geometric stratification of accounting data. Contaduría y Administración **214** (2004)
4. Er, S.: Computational methods for optimum stratification: a review. In: Proceedings of the 58th World Statistical Congress (2011)
5. Hedlin, D.: A procedure for stratification by an extended Ekman rule. J. Off. Stat. **16**(1), 15–29 (2000)
6. Aguilar, A.G., Escamilla, I. (Coords.): Segregación urbana y espación de exclusión: Ejemplos de México y América Latina, Universidad Nacional Autónoma de México, UNAM, México: MAPorrúa (2015). http://ladupo.igg.unam.mx/portal/index.php/9-publicaciones/40-libros-2015-1
7. Camacho-Sandoval, F.: Estratificación socioespacial en la ciudad de Aguascalientes en 1990, 2000 y 2010. Carta Económica Regional **127** (2020). https://doi.org/10.32870/cer.v0i127.781
8. Johnson, R.A., Wichern, D.W.: Applied Multivariate Statistical Analysis, vol. 6. Pearson, London, UK (2014)

9. Team, R.C., Team, M.R.C., Suggests, M.A.S.S., Matrix, S.: Package Stats. The R Stats Package (2018)
10. Charrad, M., Ghazzali, N., Boiteau, V., Niknafs, A.: NbClust: an R package for determining the relevant number of clusters in a data set. J. Stat. Softw. **61**(6), 1–36 (2014). http://www.jstatsoft.org/v61/i06/
11. R Core Team.: R: A language and environment for statistical computing. R Foundation for Statistical Computing, Vienna, Austria (2021). https://www.R-project.org/
12. Banerjee, S., Carlin, B.P., Gelfand, A.E.: Hierarchical Modeling and Analysis for Spatial Data. Chapman and Hall/CRC (2003)
13. Anselin, L.: Local indicators of spatial association-LISA. Geogr. Anal. **27**(2), 93–115 (1995)
14. Garrocho, C., Campos-Alanís, J.: Réquiem por los indicadores no espaciales de segregación residencial. Papeles de población **19**(77), 269–300 (2013)
15. Cressie, N.: Statistics for Spatial Data. Wiley, New York (1993)
16. Anselin, L., Syabri, I., Kho, Y.: GeoDa: an introduction to spatial data analysis. In: Handbook of Applied Spatial Analysis, pp. 73–89. Springer, Berlin, Heidelberg (2010)
17. Instituto Nacional de Estadística y Geografía (INEGI). https://www.inegi.org.mx/programas/ccpv/2020/default.html

Challenges in Performing the Quick Counts of the National Electoral Institute in Mexico

Gerardo Orantes-Jordan and Manuel Mendoza

1 Introduction

The Quick Count is a statistical procedure whose purpose is to estimate the election results from a random sample of ballot boxes. The objective of the Quick Count organized by the INE is to reduce the uncertainty about the result and convey confidence to the voters that the organizing institute can know, with an acceptable degree of approximation, the result of the election on the same day it was held. In other countries, a pronouncement of the organizing institute over the electoral results takes days after the elections; the uncertainty induced by the lack of results may lead to political conflicts.

The National Electoral Institute (INE) or its predecessor institution Federal Electoral Institute (IFE) has organized quick counts since 1994 for the elections to President of the Republic without interruption. Since 2015, it has also been used to estimate the composition of the Chamber of Deputies and since 2016 for the results of the elections to the Governor and Head of Government of Mexico City. In addition, the most recent Quick Counts have included the voter turnout estimate.

Finally, in 2007 constitutional article 116 was amended to standardize the electoral calendars in the federal entities in Mexico to make processes more efficient and less costly for the federation. This amendment increased the number of simultaneous elections for which the INE carried out a Quick Count. For example, in 2018, nine local elections and one federal elections were organized. While in 2021, fifteen local and one federal election were organized.

G. Orantes-Jordan (✉)
Morant Consultores, Mexico City, Mexico
e-mail: datos@morant.mx

M. Mendoza
Statistics Department, Instituto Tecnológico de México (ITAM), Mexico City, Mexico
e-mail: mendoza@itam.mx

I. Antoniano-Villalobos et al. (eds.), *Interdisciplinary Statistics in Mexico*,
Springer Proceedings in Mathematics & Statistics 397,
https://doi.org/10.1007/978-3-031-12778-6_9

Due to the level of demand that an exercise like this requires, the INE brings together a group of statistical experts with whom it forms the Technical Advisory Committee for Quick Counts (COTECORA) to carry out the estimation procedures. They are in charge of the sample design, propose the estimation procedure, and process the data to obtain the inferential results.

The INE requires the inference of the electoral results to be based on estimation intervals for the voting percentages with a confidence level of 95% or, in the case of a Bayesian analysis that the estimation interval accumulates 0.95 probability. In addition, the objective is to propose procedures that produce estimates with interval lengths less than 1% with the complete sample for the federal election.

A necessary clarification is that the sample is made up of ballot boxes and not people, which adds confidence to the process since the information input is the actual result of each box and not the preference reported by a citizen through an interview. However, this condition means that information cannot be obtained before the polls close, which happens at 18:00 h, local time in Mexico if there are no voters waiting in the line.

The most important restriction of the exercise is that the results must be published on election night. The logistical challenge of reporting a random sample of the size necessary to produce reliable estimates is enormous. The INE has a considerable staff supervising the election on the field, which is in charge, among other tasks, of reporting the sample results. Despite the enormous effort made by the INE, the objective of having the complete sample at the time of the final estimate has never been achieved.

To mitigate the risk of producing estimates with incomplete samples, some members of the Committee proposed technical procedures in 2021 to incorporate the uncertainty arising from not having the initially planned sample into the estimation process. Another mechanism, used since previous editions of the Quick Count, has been developed to constantly update the estimates with partial samples and monitor both the results produced and the composition of the sample. This strategy allows to assess the estimates' stability and verify that there are no imbalances in the sample that may suggest a bias in the estimates. The information system updates its data every 5 minutes, so to fulfil the objective of monitoring the results, the code with which the procedure is executed must be fast enough to handle the estimates of all the concurrent elections in that time. For the rest of the text, we will explain the strategy used to accomplish this objective.

2 The Model

As already said, the COTECORA is in charge of both the sample design and the estimation methods. Since 2006, the Mendoza and Nieto-Barajas (2016) model has been used to produce estimates of electoral results. This section describes the most relevant aspects of the model; the details and justifications for each decision of the model can be found in [1]. For 2018 and 2021, minor adjustments were made to estimate voters' turnout as described in [2]. They are not addressed in this text since these adjustments are not computationally complex. They were not part of the changes to make the code more efficient, so it was considered that addressing them did not contribute to the purpose of the text.

In 2021, a stratified random sampling was chosen to select the ballot boxes that formed the sample for the quick count. For the federal election, the strata were defined as the 300 electoral districts existing in the country.

The model uses a Normal distribution to describe the number of votes X_{ijk} that candidate i obtains in the ballot box k of stratum j. Thus,

$$X_{ijk} \sim \mathcal{N}\left(n_{jk}\theta_{ij}, \frac{\gamma_{ij}}{n_{jk}}\right) \tag{1}$$

where n_{jk} is the number of voters eligible in that specific ballot box. In addition, θ_{ij} can be interpreted as the proportion of votes for candidate i, with respect to the number of eligible voters that could have voted in that specific ballot box whereas $\frac{\gamma_{ij}}{n_{jk}}$ is the precision of the distribution.

As discussed in Mendoza and Nieto-Barajas (2016), a Normal model is a natural choice for X_{ijk}. In fact, if X_{ijkl} is defined as a *Bernoulli* random variable whose value is one if and only if the l-th voter in the k-th ballot box of stratum j votes in favor of the candidate i, then

$$X_{ijk} = \sum_{l=1}^{n_{jk}} X_{ijkl} \tag{2}$$

Thus, X_{ijk} is a sum of *Bernoulli* variables and the *Binomial* distribution arises as an option ($X_{ijkl} \sim Bernoulli(\theta_{ij})$ and $X_{ijk} \sim Binomial(n_{jk}, \theta_{ij})$). It must be noted, however, that each ballot box has several hundreds of registered voters ($500 \leq n_{jk} \leq 750$, for every ballot box). Under these circumstances, a Normal model becomes a natural approximation to the *Binomial* and, as a first choice, the usual Normal distribution with mean $n_{jk}\theta_{ij}$ and variance $n_{jk}\,\theta_{ij}(1-\theta_{ij})$ is considered (please note that, in terms of precision, a usual parametrization for a Bayesian analysis, the second parameter is given by $\left[\theta_{ij}(1-\theta_{ij})\right]^{-1} / n_{jk}$). As a final step, it is worth noting that the underlying *Bernoulli* variables might not be independent, since the voters' list of each ballot box includes clusters of people (neighbors, friends, and families, for example) whose electoral preferences are usually correlated. In order to take this fact into account, Mendoza and Nieto-Barajas (2016) replace the precision $\left[\theta_{ij}(1-\theta_{ij})\right]^{-1} / n_{jk}$ with the expression γ_{ij} / n_{jk}, where γ_{ij} is a precision parameter that does not depend on θ_{ij} and provides the model with more flexibility to describe a greater level of uncertainty.

For the θ_{ij} and γ_{ij} parameters, it was considered that the *prior* distribution should, on the one hand, reflect the INE's position of impartiality, so it should initially give each candidate the same probability of success. On the other hand, "vague" distributions had to be chosen that allowed the data to have the most significant weight in the *posterior* distribution. Thus, the following *prior* distributions were adopted:

$$p\left(\theta_{ij}, \gamma_{ij}\right) \propto \gamma_{ij}^{-1} I\left(\gamma_{ij} > 0\right) I\left(0 < \theta_{ij} < 1\right) \tag{3}$$

The *prior* distribution of θ_{ij} reflects that there is no reason to think that the proportion of votes in favor of a candidate is in a specific area of the parametric space; every region of the same length accumulates the same probability. As for γ_{ij}, this is a usual noninformative *prior* which concentrates the highest probability in values close to zero, thereby reflecting that the uncertainty is high before receiving data.

This *prior* joint distribution is updated with the information of the sample, which, as discussed, is collected using a stratified strategy. The definition of the strata, the proportional allocation of the sample as well as the total sample size were designed to produce *posterior* probability intervals for the parameters of interest with length no greater than 1% and probability 0.95. These target values are achieved when the complete sample is available. However, previous experiences have shown that often, with 75% of the planned sample, the 0.95 probability intervals have a length of approximately 2%.

Once the sample is available, the *posterior* joint distribution of the parameters θ_{ij} and γ_{ij} is obtained as

$$\begin{aligned} p(\theta_{ij}, \gamma_{ij}|X_{ij}) \sim & \mathcal{N}(\theta_{ij}|\frac{\sum_{k=1}^{c_j} x_{ijk}}{\sum_{k=1}^{c_j} n_{jk}}, \gamma_{ij} \times \sum_{k=1}^{c_j} n_{jk}) I(0 < \theta_{ij} < 1) \\ & \times \; Ga(\gamma_{ij}|\frac{1}{2}\{\sum_{k=1}^{c_j} \frac{(x_{ijk})^2}{n_{jk}} - \frac{(\sum_{k=1}^{c_j} x_{ijk})^2}{\sum_{k=1}^{c_j} n_{jk}}\}) \end{aligned} \tag{4}$$

where X_{ij} stands for the vector that contains all the sampling information regarding the votes for the candidate i in the stratum j; c_j is the number of ballot boxes in the stratum j that are included in the sample, and x_{ijk} represents the number of votes that favor the candidate i in the ballot box k in the sample of the stratum j.

Note that this joint distribution is decomposed in two factors: the marginal *posterior* distribution of γ_{ij}, and the conditional distribution of θ_{ij} given γ_{ij}. This representation is essential for the purpose of this text since it induces a simulation procedure to describe the joint *posterior* probability. First, γ_{ij} is simulated, and then conditional on that value θ_{ij} is simulated.

The θ_{ij} and the γ_{ij} parameters are estimated at the stratum level, and their estimates are independent of the estimates of the parameters in other strata. This is the case because the stratified design collects one random sample from each stratum and these samples are independent. In any case, inferences must be produced at the national level. To this end, as an intermediate step, θ_i is computed as the sum of the parameters θ_{ij} weighted by the proportions of register voters in the corresponding stratum. Thus,

$$\theta_i = \sum_{j=1}^{N} \frac{n_j}{n} \theta_{ij} \tag{5}$$

Now, the parameter we want to estimate is not the number of votes divided by the number of people eligible to vote. The ratio to be estimated is the number of votes in favor of a candidate out of the total number of casted votes. For that purpose, in the case of candidate i, the parameter of interest λ_i is defined as

$$\lambda_i = \frac{\theta_i}{\sum_{j=1}^{I} \theta_j} \tag{6}$$

where I is the number of options that each voter had for his/her vote. This strategy for the estimation of the parameters of interest has proved to be an advantage of the method proposed by Mendoza and Nieto-Barajas. As can be seen, the intermediate proportions $\theta_1, \theta_2, \ldots, \theta_I$ are estimated in a marginal fashion even though they are negatively correlated (a vote in favor of a candidate is a vote that other candidates cannot obtain). The same idea applies to the parameters of interest $\lambda_1, \lambda_2, \ldots, \lambda_I$ and there are other methods in the literature that use a multivariate structure to take into account the correlations among these proportions. Usually, inferences with these models are more complex and time-consuming than the proposal described in this paper. The idea of Mendoza and Nieto-Barajas is that producing marginal estimates for $\theta_1, \theta_2, \ldots, \theta_I$ is relatively simple (and fast) and that Eq. 6 guarantees that $\lambda_1, \lambda_2, \ldots, \lambda_I$ are negatively correlated (as expected) since they have the structure of compositional variables.

3 The Code

The first time this model was used for the Quick Count was in 2006. On that occasion, the only results that were estimated were those of the Presidency of the Republic. The execution of the code lasted, at that time, about four and a half minutes, which was enough to fulfill the time restriction. However, to use the model with simultaneous elections, it was necessary to decrease the execution time.

The first strategy that came up was to divide the primary function into processes. Before writing the code, it was necessary to propose a workflow detailing the steps of the estimation algorithm.

As a first step, a code was made that implemented the proposed algorithm without dedicating too much effort to efficiency. This code, which we will call functional code, served as a reference to test the performance of the improvements.

This functional code was written following some strategies to make it easy to use for future enhancements. In principle, a recommendation is to follow a consistent code style. This makes collaborative work more manageable and makes it easy to replace a piece of code with an improved version. In our case, we stuck to [3], since the packages used, for the most part, are part of the tidyverse.

There are some standard practices to reduce execution time, for example, avoiding repeating calculations. If there is a recurring calculation in the procedure, it is better to store the variable and reduce the execution time. Another standard procedure is to vectorize functions whenever possible. That is, to generalize the functions so that instead of iterating through "for loops", they apply directly to the array of interest. On this point, it should be clarified that the iterations through "for loops" are not as slow as they used to be, but avoiding them represents a gain in code clarity.

Once a functional code is achieved, performing a function performance analysis procedure known as Profiling is convenient. This procedure helps identify the memory used by each function in the code and also the execution time. There are different methods for Profiling. Newer methods use a sample of chunks of code that they execute and record the time they take and the memory they occupy. In R, the profvis package provides a friendly graphical interface to perform the Profiling of a function and to identify which processes occupy more memory and consume more time.

After the initial diagnoses, it is crucial to focus on the sub-processes that are more time-consuming. An element that is sometimes overlooked when trying to produce an efficient code is that before coding, there may be a way to pose a problem that facilitates the implementation of the solution.

Without a doubt, the main advantage of open software, in particular of R, is collaborative work. After having a functional code with which to compare the results, it can be inquired if there are packages that contain functions that make a part of our code faster. Going to the community is usually very useful if the results can be contrasted against a functional code of our own.

Finally, when trying to reduce code execution times, it is vital to consider the effort involved in making the code faster and contrast it with the execution time constraints. An estimation can be implemented faster in almost all cases, but it is crucial to see if a simple solution can be provided before attempting complex solutions.

3.1 The Functional Code

In the case of the implementation of the Mendoza and Nieto-Barajas (2016) model, we identified that the algorithm to produce an estimate of the electoral results could be described in four steps:

1. Read the database.
2. Calculate the required summaries for each stratum and each candidate.
3. Simulate the γ_{ij} and θ_{ij} parameters for each stratum and each candidate.
4. Calculate the λ_i parameters.

On the day of the election, some additional steps associated with the day's operation were carried out. For example, some procedures to verify that the database had the expected format: to make sure the sample did not contain invalid data, that the new accumulated database contains the information of the previous databases, and that the ballot boxes used were part of the sampling frame. Some other sub-processes were also included to visualize the results of the election. None of these procedures is necessary to produce the requested results, so they were not dealt with in this text.

For each of the steps of the algorithm, a function was constructed. Three of the four functions do not represent any complexity from a computational point of view.

Read the base: The database was delivered in a very light format; no document exceeded 1,000 KB. There are several functions for reading text in tabular format, a contrast of different functions can be found in [4]. However, the execution time is so low that any choice of function to read the base will not have a significant impact on the execution time.

Calculate the required summaries for each stratum and each candidate: To obtain these values, that act as parameters of the distribution of γ_{ij} and θ_{ij}, we need to perform simple arithmetic operations, multiplication, addition, subtraction, and division. These operations have a low computational cost.

For our case, we store the data in an object of class tibble. These objects are the preferred way to work with the tidyverse. It is an object for tabular databases that is more effective than data.frames, and more expressive for the user than an array. In addition, it can have lists as columns, so complex objects such as graphs or simulation vectors can be stored in a cell. Details of the tibbles can be found in [5]. There are many ways to calculate the values for each stratum. We use the approach based on direct applications of the formula using the group_by, mutate, and summarize functions from the dplyr package. In this way, we end up with a new database whose elements contain data of candidates for each stratum.

Simulate θ_{ij} and γ_{ij} by candidate and stratum: This step is the only one that represents complexity from a computational point of view. Since the first implementation of the model in 2006, it was observed that 10,000 simulations per stratum and candidate were sufficient to describe the *posterior* probability distributions. Therefore, in principle, the number of simulations is the result of multiplying 10,000 by the number of strata and by the number of candidates. In the case of the 2018 federal election, the number of simulations was 35,000,000 every five minutes. The large number of simulations needed implies that a saving of time per simulation, even if it is minimal, can considerably impact the overall execution time.

The distribution that interests us is the marginal distribution of θ_{ij}; it can be obtained from the joint distribution of both parameters and in that sense, γ_{ij} is an auxiliary parameter that helps us describe θ_{ij}. The precision parameter can be simulated directly from a Gamma distribution (see Eq. 4). Then, we can simulate θ_{ij} conditional on the value of γ_{ij}; if θ_{ij} falls within the interval [0, 1], then it holds, if it does not enter the interval, the simulation is repeated. This way of proceeding implies a rejection criterion, that is to say, that not all simulations are used, some are discarded, which potentially increases the number of simulations. The smaller the γ_{ij}, the precision, the higher the probability of rejection.

This mechanism is programmed with the described rejection criteria. The purrr::pmap function was used to generalize the procedure without the need for an explicit loop. Thus, a vector of 10,000 simulated values of θ_{ij} is obtained for each pair, stratum, and candidate. The versatility of the tibbles allows us to add as a list-column all the simulations to the database of strata and candidates that was produced in step 2. In addition, it facilitates the calculations of the national parameters by organizing the simulation vectors with the rest of the data from the strata and candidates.

Calculate λ_i: Before calculating λ_i, an intermediate step calculates θ_i at the national level. To this end, it is enough to make sums weighted by the size over all θ_{ij} parameters of the corresponding candidate (see Eq. 5). Once again, the tidyverse tools could be articulated to facilitate the process; using the tidyr::unnest function, it is possible to use the dplyr functions again but this time to calculate the values of the national parameters, both θ_i and λ_i. This process is reduced once again to arithmetic operations of low computational complexity. In short, the estimation algorithm, in general, has computationally inexpensive threads. However, the simulation of γ_{ij} and θ_{ij} offers possibilities to reduce execution time.

Figure 1 shows the execution time for each database of the 2018 federal election. Note that, in general, we obtain sufficiently short execution times with the strategies followed in the functional code. It is important to note two issues. The first is that the execution time depends on the database. As information accumulates, the execution time decreases. The reason for the time reduction in the latter databases is that as there is less uncertainty than at the beginning of the process, the probability that a simulation of θ_{ij} is in the rejection region is lower. Another critical issue is to notice the variability in the execution time. The analysis of the graph of execution times induces the objective of reducing the time in a general way, but above all, to control the variability so as not to run the risk of having executions that exceed the desirable execution time. Even with a large sample, the risk of having a slow estimation is present as we can see in Fig. 1. This is possible because of the existence of a rejection region. The general trend is to get lower execution times as the sample size grows, however, there is always a chance of getting a slow execution time because of having to repeat the sampling when the value of θ_{ij} is not in the domain.

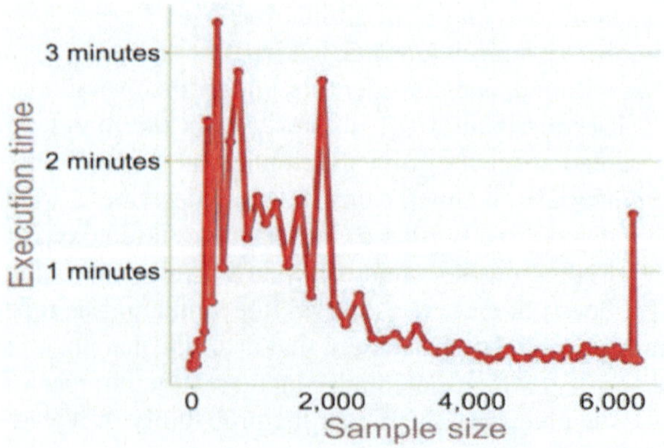

Fig. 1 Execution time for the functional code by sample size of the corresponding database

3.2 First Idea: Use the Sample Design

Until now, the functional code consists of a few lines arranged in four steps, of which only a step takes the most amount of time. Two critical aspects stand out in the step of the simulation of γ_{ij} and θ_{ij}. The first is the number of simulations. The second is the rejection criterion. On the number of simulations, we observe that it depends on the number of strata, the number of candidates, and the constant, in this case, 10,000, which is necessary to obtain a clear description of the probability distribution. The number of candidates is fixed and cannot be changed. On the other hand, reducing the number of simulations is detrimental to the quality of the inference. The number of strata, meanwhile, is a decision that corresponds to the sampling design which is not appropriate to determine based on computational efficiency.

However, one element that can be taken advantage of is the fact that inferences on the parameters of interest θ_{ij} and γ_{ij} are independent at the stratum level. Therefore, it is unnecessary to update estimates for all strata, but only for those strata for which there is new information on the latest arrival

1. Read the current database.
2. Calculate the interest values for each stratum.
3. Filter strata with new information.
4. Simulate the γ_{ij} and θ_{ij} parameters for strata with new information.
5. Join the strata with the new information database and previous strata database.
6. Calculate the λ_i parameters.

Implementing this idea results in an impact on the estimation algorithm. It adds two new steps and increases the number of parameters passed to the function. These changes may involve, in principle, a computational cost. However, it can be countered with the benefit of reducing the number of simulations.

It is observed that one of the objectives is met, that of reducing the execution time, in general (see Fig. 2). However, controlling for variability in execution time remains a pending issue. In almost all cases, the time is less than one minute; however, the time exceeds two minutes in some iterations.

3.3 Second Idea: Reach Out to the Community

One of R software's greatest strengths is its community. This community grows every day and increases the power of R day by day through the sharing of functions contained in libraries. Once we have a functional code that executes the desired estimation, and after analyzing its operation and identifying that the most expensive step is the simulation of a Truncated Normal, we can go to the R community to see if there is an efficient implementation of Truncated Normal simulations.

The truncnorm package contains functions to compute the density, the cumulative probability, and the quantiles of a Truncated Normal distribution. It also provides a

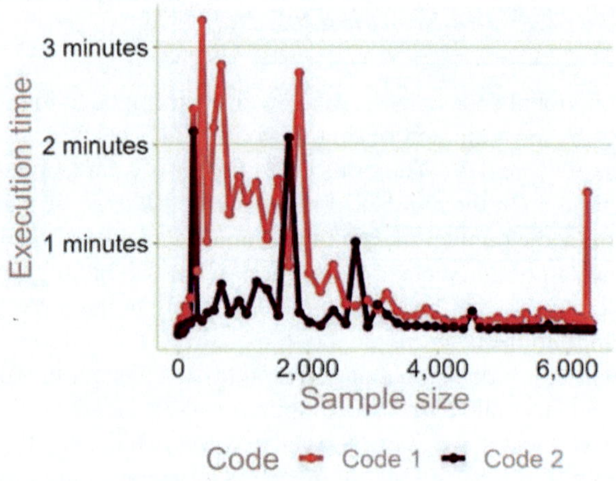

Fig. 2 Execution time of functional code (Code 1) and the first modification (Code 2) by the sample size of the corresponding database

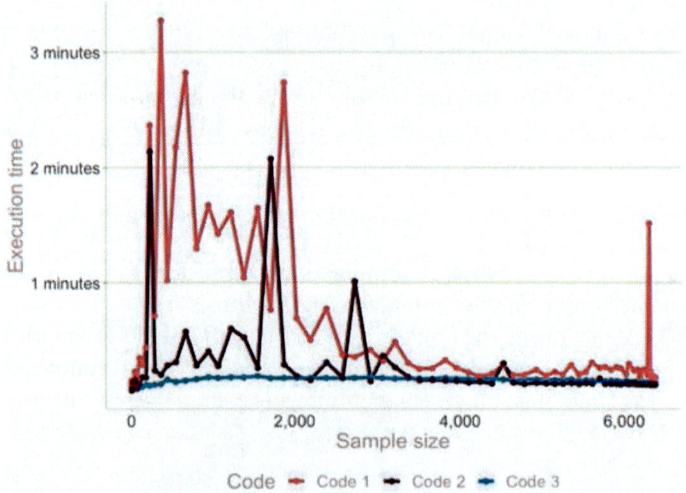

Fig. 3 Execution time of functional code (Code 1), first (Code 2), and second modification (Code 3) by the sample size of the corresponding database

function to generate random numbers from this model. In particular, the rtruncnorm function generates numbers efficiently for two reasons; it uses an efficient rejection criterion proposed by [6], and the implementation is not done in R but uses the lower-level C/C++ language. These languages usually require higher specifications from the programmer; however, they make code execution faster. This modification does not have an impact on the structure of the code; it is enough to replace the function that generates the random numbers. Thus, using the truncnorm package the objectives of reducing, in general, the execution time was achieved (see Figs. 3

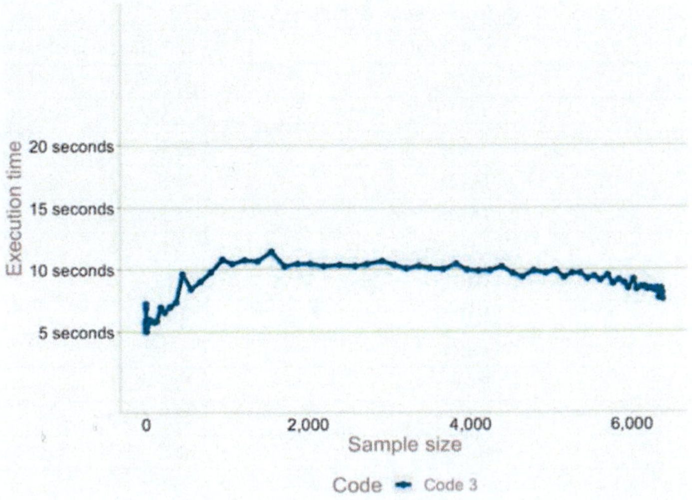

Fig. 4 Execution time of the second modification by the sample size of the corresponding database

and 4). In addition, the variability of the execution times is controlled. Lastly, the execution time of the estimation code never exceeds 15 s for the national election. Local elections have even fewer strata, so the goal of reducing the time of execution of the code to deal with many simultaneous elections was also reached.

Finally, although it should not happen, code changes could produce different estimation results. Table 1 shows the last estimate produced under the different strategies. It can be seen that the differences are minimal, and in no case are they relevant for the analysis. Thus, the strategies do not alter the estimates but bring benefits in the execution times.

Other strategies could be followed to decrease execution time further; for example, we may use languages that are faster than R and try other implementations of random number generation from a Truncated Normal distribution such as [6, 7]. However, the function that fulfills the Quick Count time constraints is loose, so we feel that, with the proposed modifications, the objective has been achieved.

4 Concluding Remarks

In many real statistical applications, the inferential methods must be optimal, and the specific results must be produced in a reasonable period of time. This is the case of the quick counts, whose final inferences should be available very soon after the last database is available.

In addition, if the statistical analysis relies on a relatively heavy simulation process and the quality of the final results requires not just the analysis of the last database

Table 1 Last estimation with different codes. Sample size was 6,381

Code	Candidate	Confidence interval (%)
Code 1	AMLO	52.9–53.6
Code 2	AMLO	52.9–53.6
Code 3	AMLO	52.9–53.6
Code 1	JAMK	15.8–16.4
Code 2	JAMK	15.8–16.4
Code 3	JAMK	15.8–16.4
Code 1	JHRC	5.3–5.6
Code 2	JHRC	5.3–5.6
Code 3	JHRC	5.3–5.6
Code 1	RAC	22.0–22.5
Code 2	RAC	22.0–22.5
Code 3	RAC	22.0–22.5

but to apply the entire process to a sequence of increasing databases, the need for an efficient code becomes an issue.

In this paper, we have shown how, within the environment of the R language and using some basic ideas, the execution time for the INE quick count can be stabilized and reduced from maximums above three minutes to around ten seconds. Of course, even a greater reduction might be achieved if a lower-level language is used instead of R, and this possibility could be explored in future elections.

Acknowledgements Partial support from the Sistema Nacional de Investigadores (Mexico) is gratefully acknowledged. M. Mendoza wishes to acknowledge support from the Asociación Mexicana de Cultura A.C.

References

1. Mendoza, M., Nieto-Barajas, L.E.: Quick counts in the Mexican presidential elections: a Bayesian approach. Elect. Stud. **49**, 124–132 (2016)
2. Orantes-Jordan, G.: Conteo Rápido para la elección presidencial de 2018 en México: Análisis Bayesiano. ITAM Thesis (2019)
3. Wickham, H.: The tidyverse style guide. https://style.tidyverse.org/
4. Colin G., Robin L.: Efficient R Programming. O'Reilly Media (2016)
5. Grolemund, G., Wickham, H.: R for Data Science. O'Reilly Media (2017)
6. Geweke, J.: Efficient simulation from the multivariate normal and student-t distributions subject to linear constraints and the evaluation of constraint probabilities. In: Computing Science and Statistics: Proceedings of the 23rd Symposium on the Interface, pp. 571–578. Interface Foundation of North America, Inc., Fairfax, Virginia (1991)
7. Chopin, N.: Fast simulation of truncated Gaussian distributions. Stat. Comput. **21**(2), 275–288 (2011)

Sampling Design and Poststratification to Correct Lack of Information in Bayesian Quick Counts

Carlos S. Pérez-Pérez and Luis E. Nieto-Barajas

1 Introduction

In Mexico, public organizations produce exit polls and quick counts on the election day to estimate the final result and avoid uncertainty while the complete counting of votes is done. Throughout this work, Bayesian statistical procedures are used to estimate trends of the final results of the federal election from a probabilistic sample in a timely manner. We use a Bayesian parametric model which relies on information collected from the polling stations in a random sample. This model was used by the National Electoral Institute (INE) during the 2006, 2012 and 2018 presidential elections and in some governor state elections in Mexico. However, the 2021 federal election aimed to define the composition of the chamber of deputies (COD).

The COD in Mexico is formed by 500 representatives: 300 of them are elected by direct voting to represent their district, whereas the remaining 200 are determined by proportional representation of the overall votes in the whole country.[1] The problem of estimating the composition of the chamber, from a statistical point of view, requires estimations at two levels: first, at the level of each of the 300 federal districts, for the deputies elected by simple majority; and secondly, at the national level for the designation of the 200 deputies by proportional representation. The former requires a sample representation at the federal district level and the latter a representation at the national level. The most appropriate sample design for the problem in question is therefore a stratified random sampling by federal district level with the particularity that strata are also study domains. Election regulations establish the principles for

[1]By federal law, parties with less than 3% of valid voting, null plus unregistered and independent have to be excluded from the distribution of seats.

C. S. Pérez-Pérez (✉) · L. E. Nieto-Barajas
Department of Statistics, ITAM, Mexico City, Mexico
e-mail: carlos.perez@itam.mx

L. E. Nieto-Barajas
e-mail: lnieto@itam.mx

I. Antoniano-Villalobos et al. (eds.), *Interdisciplinary Statistics in Mexico*,
Springer Proceedings in Mathematics & Statistics 397,
https://doi.org/10.1007/978-3-031-12778-6_10

the determination of the deputies elected by relative majority, and the principles for the election of deputies by proportional representation.

In this article, we present the details for the definition of the sampling design and introduce a poststratification criteria to deal with incomplete samples in the estimation of the percentage of valid votes and the number of members per political party or independent candidacies that are chosen to constitute the COD. Poststratification is a technique for adjusting a non-representative sample after the sample has been selected, so it tackles among many issues, missing information [1]. This scientific procedure can be combined with regression, providing estimates based on features rather than raw counts in each stratum [2]. As described in [3], multilevel modeling is useful in determining how much partial pooling to apply in the regressions, leading to the popularity of the combination of multilevel regression and poststratification (MRP) [4]. A recent application of multilevel regression in the context of the quick counts [5] uses a Bayesian hierarchical model to produce estimates for Mexican governorship elections. They consider poll stations poststratified by demographic, geographic, and other covariates, to deal with biases associated with the information arrival pattern.

In this volume, [6] presents another interesting way of dealing with the lack of information in quick counts. In particular, he uses multiple imputation techniques.

In Sect. 2, we define the parameters of interest, frame the problem as an estimation task and present the model specifications. In Sect. 3, we present the sampling design based on resampling techniques using results from previous elections to quantify the uncertainty. In Sect. 4, we introduce the poststratification via hierarchical clustering, as well as the procedures for incomplete sample estimation. Results of the election day are shown in Sect. 5, whereas conclusions and further work are discussed in Sect. 6.

2 Model and Specifications

Regulations and criteria to determine the conformation of the COD are detailed in [7]. Although the criteria are based on the total votes per party, it is possible to represent it in terms of the proportions of votes, which we introduce in the following notation. In the remainder, we will use the term "nominal list"; this consists in the registry of voters who can potentially attend the polling stations during the election day.

Let $\theta_{i,j}$ be the proportion of votes, relative to the nominal list, in favor of the candidate/party j in the stratum (district) i, with $j = 1, \ldots, J$ and $i = 1, \ldots, N$. J is the total number of candidates (which includes candidates by political parties and independents) plus two extra categories: null plus unregistered votes and abstentions. $N = 300$ is the number of federal districts. Let n_i be the size of nominal list of voters in district i for $i = 1, \ldots, N$, and $n = \sum_{i=1}^{N} n_i$ the size of the national nominal list. Then $\theta_j = \sum_{i=1}^{N} \frac{n_i}{n} \theta_{i,j}$ is the proportion of votes, relative to the nominal list, in favor of candidate j at the national level.

For the determination of the winning deputies by *relative majority*, we define

$$\text{RM}_{i,j} = I(\theta_{i,j} > \theta_{i,l}); \forall l \neq j, l = 1, \dots, J-2$$

an indicator that tells us whether or not candidate j won in district i.

Finally, we define $\text{RM}_j = \sum_{i=1}^{N} \text{RM}_{i,j}$ as the number of deputations won by the candidate j by the principle of relative majority.

To determine the winning deputies by *proportional representation*, we first define $\lambda_j = \theta_j / \sum_{l=1}^{J-2} \theta_l$ the proportion of valid votes, where abstentions are excluded, and $\eta_j = \theta_j / \sum_{l=1}^{J-3} \theta_l I(\lambda_j > 0.03)$ as the proportion of votes after excluding parties that do not reach 3% in valid voting (i.e., $\lambda_j \leq 0.03$), independents and null plus unregistered.

The proportional representation deputies are determined via an iterative algorithm that consists in first calculating $\text{PR}_j = \lfloor 200\eta_j \rfloor$ the seats by raw proportional representation for the party j and then adjusting according to an over-representation constraint, i.e., if $\text{RM}_j + \text{PR}_j > \lfloor 500(\eta_j + 0.08) \rfloor$ so $\text{PR}_j^* = \lfloor 500(\eta_j + 0.08) \rfloor - \text{RM}_j$. Subsequently, the remaining decimals $200\eta_j - \text{PR}_j$ are calculated and ordered from highest to lowest and an additional deputation is assigned to each party until completing 200.

Finally, the number of seats per political party NS_j results from adding the seats by relative majority RM_j and those of proportional representation PR_j^*. That is, $\text{NS}_j = \text{RM}_j + \text{PR}_j^*$. For a more detailed explanation of the deputies assignment algorithm, we refer the reader to [6] Appendix in this volume.

For the estimation of the parameters described above, and in particular to estimate $\theta_{i,j}$, we used a Bayesian approach. Specifically, we used a slight modification of the model proposed by [8]. This model assumes that $X_{i,j,k}$ denotes the votes in favor of the party/candidate j in the stratum i in the polling station k. This observable variable follows a behavior described by the following distribution

$$X_{i,j,k} | \theta_{i,j}, \tau_{i,j} \sim \text{N}\left(n_{i,k}\theta_{i,j}, \frac{\tau_{i,j}}{n_{i,k}}\right)$$

where $k = 1, \dots, c_i$ with c_i the sample size of polling stations in the district i, $i = 1, \dots, N$ and $j = 1, \dots, J$. Additionally, $n_{i,k}$ is the size of the nominal list for the polling station k in the district i and $\tau_{i,j}$ is a precision, assumed constant within the stratum and independent of $\theta_{i,j}$. Furthermore, $X_{i,j,k}$ is assumed independent of $X_{i,j',k}$ for $j \neq j'$.

The original prior distribution considered in [8] assumes independence and is of the form $\theta_{i,j} \sim \text{U}(0, 1)$ and an improper prior $p(\tau_{i,j}) \propto \tau_{i,j}^{-1} I(\tau_{i,j} > 0)$, where $I(A)$ denotes the indicator function of set A. However, this prior only produces a proper posterior if sample size $c_i \geq 2$, wasting valuable information when we only have a single polling station. To overcome this problem, we now propose to use a proper but vague prior of the form $\tau_{i,j} \sim \text{Ga}(0.5, 0.05)$, which has a large variance of 200.

Thus, using Bayes' theorem, the posterior distribution turns out to be proportional to the product of a truncated normal for $\theta_{i,j}$ conditional on $\tau_{i,j}$ and a gamma distribution for $\tau_{i,j}$. Specifically, we have

$$f(\theta_{i,j}, \tau_{i,j} \mid \text{data}) \propto \mathrm{N}\left(\theta_{i,j} \,\middle|\, \frac{\sum_{k=1}^{c_i} x_{i,j,k}}{\sum_{k=1}^{c_i} n_{i,k}},\ \tau_{i,j}\sum_{k=1}^{c_i} n_{i,k}\right) I(0 < \theta_{i,j} < 1)$$
$$\times\, \mathrm{Ga}\left(\tau_{i,j} \,\middle|\, \frac{c_i}{2}, \frac{1}{2}\left\{\frac{1}{10} + \sum_{k=1}^{c_i} \frac{x_{i,j,k}^2}{n_{i,k}} - \frac{\left(\sum_{k=1}^{c_i} x_{i,j,k}\right)^2}{\sum_{k=1}^{c_i} n_{i,k}}\right\}\right) \quad (1)$$

Posterior point and interval estimates for $\theta_{i,j}$ are approximated via simulations from the posterior distribution. This is required due to the truncation. For more details, see [8].

3 Sample Design

Stratification is a technique developed for sampling designs in which a population is partitioned into subgroups (strata) and each group (stratum) is sampled independently. If the subgroups are more homogeneous (i.e., lower variance) than the population as a whole, variance is reduced in the estimator of a quantity of interest at the population level [9].

As mentioned previously, the sample design for the 2021 quick count is based on a stratified random sampling by federal district. To determine the sample size and the estimation errors of each size, we carried out a simulation study, using data from the 2015 federal election with the following specifications: 1,000 stratified random samples of polling stations of sizes 3,000, 4,500, 6,000, 7,500 and 9,000 were generated which, divided among the 300 electoral districts, are equivalent to 10, 15, 20, 25 and 30 polling stations per federal district.

For each sample, the parameters $\theta_{i,j}$ are estimated in the Bayesian fashion described above with 10,000 simulation draws from the posterior distribution. For each of these simulations, we calculated the number of seats NS_j. Considering an absolute value loss function, we estimated NS_j by means of the median which we denote by $\widehat{NS}_j$.

We defined the estimation error as the number of misallocated seats per party $\mathrm{NMS}_j = |\dot{\mathrm{NS}}_j - \widehat{\mathrm{NS}}_j|$, where $\dot{\mathrm{NS}}_j$ is the official value of seats obtained by the j party described in the official page [10]. Results of the simulation exercise are shown in Fig. 1, where we present histograms of the 1,000 values of NMS_j, $j = 1\ldots, 11$ using data from the 2015 federal election.

We note that the party with the greatest variability is PAN, with errors of up to 10 seats, followed by PRD and MORENA, with errors of up to 7 and 5 seats, respectively, for a sample of size 3,000. The other parties show very little variability with practically zero errors. Comparing the sampling distribution of the errors per party in the different sample sizes, a slight increase in the variability of the errors is observed for the sample size of 3,000 polling stations (10 polling stations per district) as com-

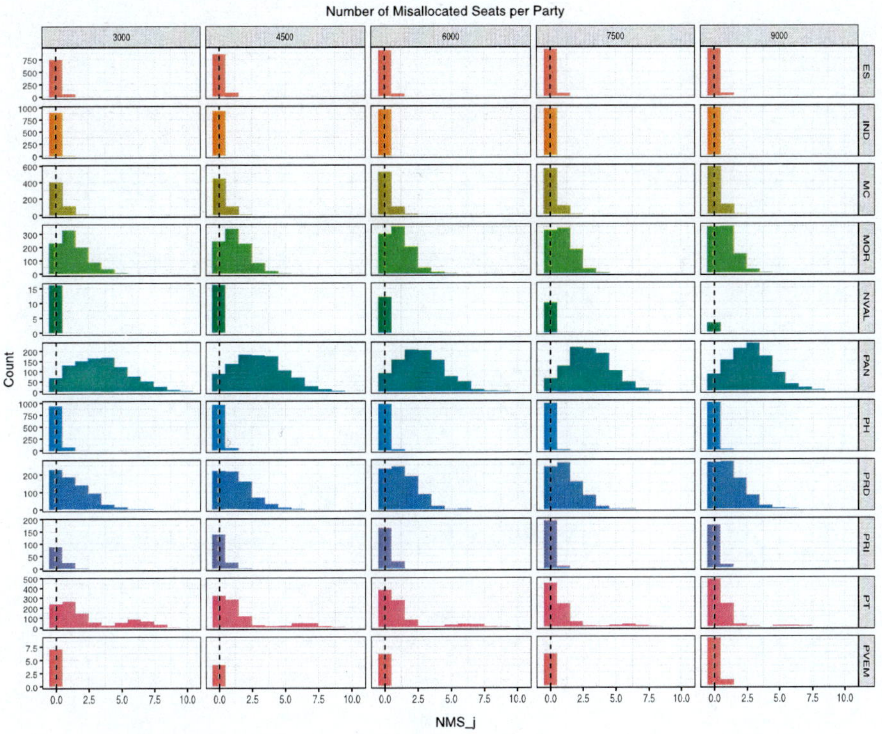

Fig. 1 Sampling distributions for NMS_j, $j = 1, \ldots, 11$

pared to larger sample sizes. An important clarification here is that the labor party (PT), corresponding to $j = 5$, is on the verge of losing its registration, in fact the true value after counting all votes is $\lambda_j = 0.029$. This means that for some samples and for some realizations of the posterior distribution of λ_5, PT may not have seats assigned by the proportional representation principle. Consequently, the NMS_j indicators for some parties such as PAN, MORENA and PRD, which are the largest parties, take larger error values compared to other parties.

We summarize the estimation error in the conformation of the entire chamber using the average number of misallocated seats $\text{ANMS} = \frac{1}{J-2}\sum_{j=1}^{J-2}\text{NMS}_j$. The sampling distribution of this indicator is included in Fig. 2 for the different sample sizes considered. In these graphs, a change in the dispersion of the distributions is observed as the sample size increases. The sampling distribution with 3,000 polling stations is much flatter (more dispersed) than the one obtained with 9,000 polling stations, which is highly concentrated on the value of one. Comparing the modes of the distributions, from 4,500 and up to 9,000, the mode is at one misallocated seat. On the other hand, the mode when we have samples of 3,000 polling stations is shared by the values of one and two error seats.

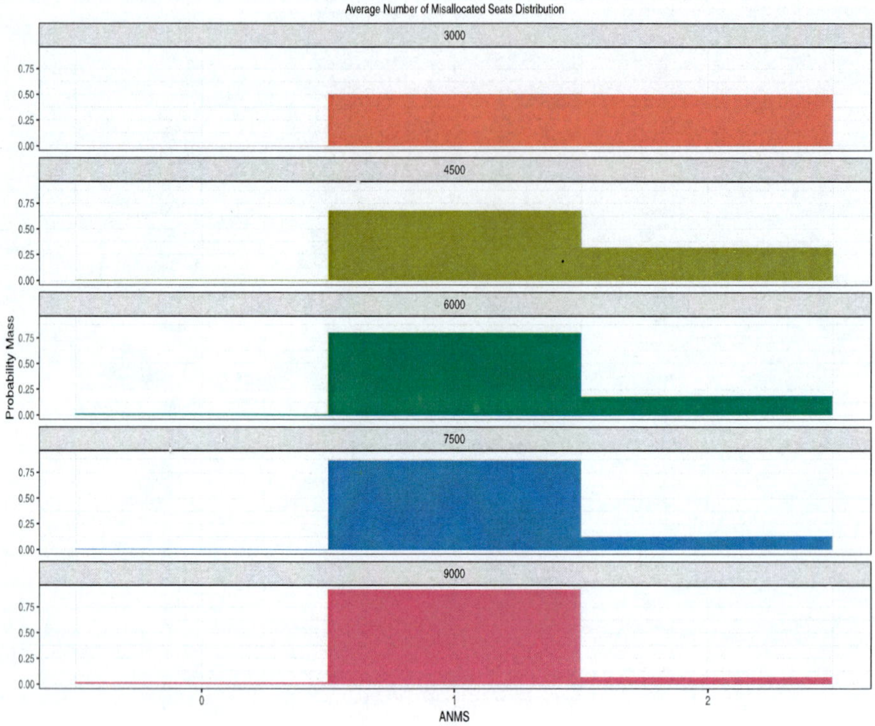

Fig. 2 Sampling distributions for $ANMS$

In summary, it was determined that our stratified sampling design, based on federal districts and a sample size of 6,000 polling stations achieves a maximum margin of error of 6 misallocated seats and an average error of 2 seats per party with a credibility level of 95%.

4 Incomplete Sample Estimation

4.1 General Strategy

On election day, data arrive gradually as they become available. Partial data constitute an incomplete sample where it is possible that some $X_{i,j,k}$ are not available for a whole stratum i. We cannot wait for the full sample, as that might never happen due to several reasons: a polling station in the sample was not set up, the person in charge of collecting the data did not attend the polling station in time, inconsistencies in the database, etc. Therefore, we have to produce estimates with the incomplete samples.

We describe here a proposal to estimate the conformation of the COD under this scenario.

We need to adjust the non-representative sample available, so we build another stratification criteria, different from the sampling design stratification that helps us solve the lack of information. This is also known as poststratification [1]. Our poststratification consists in clustering the original sampling design strata (electoral districts) into larger groups. We define a set of several poststratifications that ranges from the trivial one P_1, which contains all strata in a single group to the most granular configuration P_N, which coincides with the original strata with $N = 300$ groups. In particular, we only consider seven clusterings P_l of sizes $l \in \{1, 10, 20, 50, 100, 200, 300\}$.

Bearing in mind that we have to produce estimates at the original strata level and at the national level; for inference purposes, we maintain the original stratification of size 300 districts, with the following changes for strata with no sample available. Let $P_l^{(i)}$ be the specific group of clustering P_l that contains district $i = 1, \ldots, 300$. Let i be one of the 300 strata for which there is no information $X_{i,j,k}$ for any $k = 1 \ldots, c_i$ of the polling stations in the sample. Let l^* be the largest l for which $P_l^{(i)}$ has sample information available. Let N_{i,l^*} be the number of districts with sample information in $P_{l^*}^{(i)}$. According to posterior distribution (1), the sufficient statistics for $(\theta_{i,j}, \tau_{i,j})$ are $T_{i,j}(\mathbf{X}) = \sum_{k=1}^{c_i} X_{i,j,k}$ and $U_{i,j}(\mathbf{X}) = \sum_{k=1}^{c_i} \frac{X_{i,j,k}^2}{n_{i,k}}$. Additionally, we define $v_{i,j} = \sum_{k=1}^{c_i} n_{i,k}$. We consider the poststratification group $P_{l^*}^{(i)}$, average the sufficient statistics of the districts with sample and impute the sufficient statistic for the district without data assuming that the statistic is the result of observing a single polling station with maximum nominal list of voters. In notation, we take

$$T_{i,j} = n_0 \frac{\sum_{i'}^{N_{i,l^*}} T_{i',j}}{\sum_{i'}^{N_{i,l^*}} v_{i',j}} \quad \text{and} \quad U_{i,j} = n_0 \frac{\sum_{i'}^{N_{i,l^*}} U_{i',j}}{\sum_{i'}^{N_{i,l^*}} v_{i',j}}$$

with $n_0 = 750$ and set $c_i = 1$ and $n_{i,k} = n_0$ for inference purposes.

The sampling design, described in Sect. 3, was created to provide interval estimates with 95% credibility level. This is only achieved with the complete sample. With incomplete samples and even with the sufficient statistics imputations proposed, posterior intervals would not be credible at the nominal level. We propose to adjust the credibility level according to the percentage of sample available. We start producing estimates at a 99% credible level if the percentage of sample received is low, and gradually move the credible level, as more sample becomes available, until we reach the desired 95% credible level.

Details of how to create the set of poststratifications as well as the correction in the credibility level are given below.

4.2 Poststratification

Hierarchical clustering is a method of cluster analysis which seeks to build a hierarchy of groups. Strategies for hierarchical clustering generally fall into two types: agglomerative (from many to one, or bottom-up) or divisive (from one to many, or top-down). Independent of the approach, a measure of similarity between sets of observations is required, and this is achieved by an appropriate metric $d : \mathbb{R}^p \to \mathbb{R}_+$ and a linkage criterion as a function of pairwise distances of observations in the sets.

Most common linkage criteria include Ward's minimum variance method that aims at finding compact and spherical clusters, the single linkage method adopts a "friends of friends" clustering strategy, the complete linkage method relies on furthest neighbors and the average method makes a compromise between the last two. The results of the hierarchical clustering analysis are often represented in a dendrogram, which is a diagram depicting a tree structure that illustrates the different arrangements of the individuals into groups, as can be read with further detail in [11].

The electoral preferences in the two most recent federal elections, 2015 and 2018, were used to characterize the federal districts in a p-dimensional vector. The metric d used for the clustering process was the standardized Euclidean distance. Information from the two elections was considered separately and produced two different sets of hierarchical clusterings, using different linkage criteria.

To compare the clusterings from the two elections, we use a scoring function that measures the similarity between the two corresponding tree structures from the 2015 and 2018 data and for the different linkage criteria. This enables us to quantify the cohesion of electoral preferences, and therefore, to choose the criteria that produce the most consistent clustering configurations between election years.

In particular, we use a function ρ that quantifies the similarity between trees, say T_{2015} and T_{2018}, as the proportion of nodes (groups) in common. This and other alternatives can be found in [12]. We choose the best linkage criteria by maximizing this score. Table 1 presents the values of ρ obtained.

Since the values of ρ obtained with Ward and complete linkages are the same, we chose T_{2015} with complete linkage to build the poststratification.

Note that with the resulting tree structure T_{2015}, we can always choose a configuration P_l of exactly l groups and calculate the proportion of votes in favor of each candidate. The voting proportions for each combination of group and political party

Table 1 Score function for the different linkage criteria. If $a_i \in A, b_j \in B$, and $S(C) = \sum_{i=1}^{n_C} |a_i - \frac{1}{n_C}\sum_{j=1}^{n_C} c_j|^2$ and AB stands for the combined cluster from the fusion of A and B

Linkage	Formula	Score ρ
Single	$\min d(a_i, b_i)$	0.5091
Average	$\frac{1}{N_A * N_B}\sum_{i=1}^{n_A}\sum_{j=1}^{n_B} d(a_i, b_j)$	0.5075
Ward	$S(AB) - [S(A) + S(B)]$	0.5141
Complete	$\max d(a_i, b_j)$	0.5141

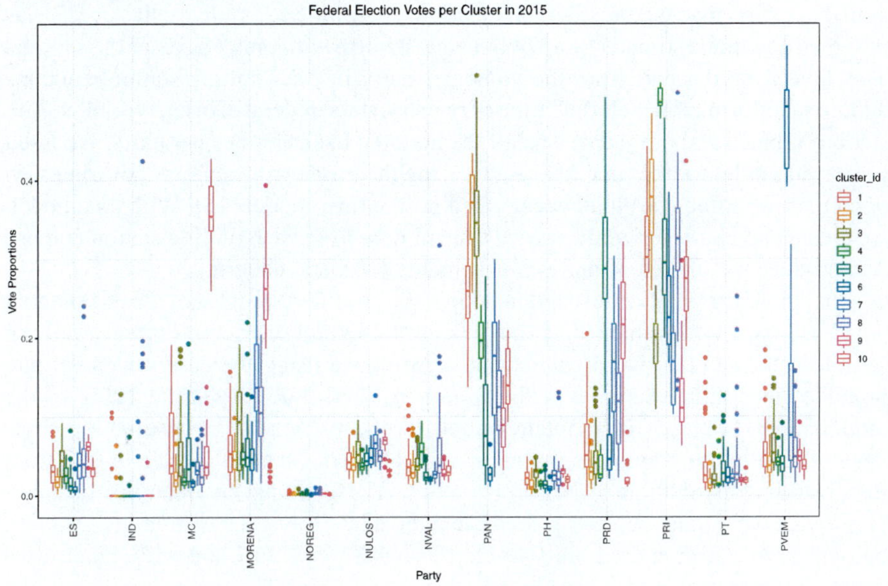

Fig. 3 Boxplots for the proportion of votes for each candidate in each of the 10 groups of P_{10}

for the P_{10} configuration are shown in Fig. 3. We can see that the political preferences are somewhat homogeneous inside each group and heterogeneous across different groups.

The hierarchical clustering and tree-correlation computations were performed via the functions in the stats library from statistical software R [13]. The Bayesian model was programmed from scratch in R and all graphics and data manipulation were done using the tidyverse suite [14].

4.3 *Credibility Level Correction*

As mentioned before, inference obtained with incomplete samples might not produce posterior intervals at the desired nominal 95% level. To determine the best credibility level correction, we implemented a simulation study. The previous two federal elections for the COD are those of 2015 and 2018. The 2015 election has the particularity that one party (PT) is on the verge of losing its registry because $\lambda_5 = 0.029$ is less than 3%, so it should not have received seats according to the proportional representation principle. However, in reality, it did receive seats, producing a disagreement between observed data and final numbers in the conformation of the COD. For this reason, we decided to work with the 2018 federal election for our study.

Although people voted for the conformation of the COD in the 2018 federal election, there was no quick count for estimating the COD. The 2018 quick count aimed at estimating the presidential election. A stratified sample of size 7,787 was designed and final estimates were obtained at 22:30 h with only 67.5% of the sample. This final dataset, apart from the votes per party in each polling station, contains additional information including time of registry, state, federal district, type of polling station (urban, rural or mixed), size of the nominal list and votes per party. We fitted a survival model for the time of registry using these other variables as covariates. In particular, we fitted a Weibull accelerated failure time model [15]. With this model, we are able to characterize the arrival time of data from each polling station and use it to produce incomplete samples in a somehow realistic pattern.

Our simulation study consisted in taking 100 stratified samples of size 6,000, that is, 20 polling stations in each of the 300 strata (federal districts). For each polling station in the sample, we predicted the arrival time using the survival model and produced incomplete samples with 50%, 60%, 70%, 80%, 90% and 100% of the total sample. For each incomplete sample, we implemented the estimation strategy outlined in the previous subsection and obtained a sample of size 10,000 from the posterior distribution of NS_j for $j = 1, \ldots, 10$. Using the quantiles $\alpha/2$ and $(1 - \alpha/2)$, we obtain a posterior credible interval at $(1 - \alpha)100\%$ nominal level. In particular, we consider credibility levels 99%, 98%, 97% 96% and 95%. Finally, we compute the coverage of intervals at each credibility level as the percentage of the 100 replicates that contain the true number of seats.

Table 2 reports the coverage for four selected parties, PAN, PRD, PRI and PVEM. For each party, the coverage behavior is different. If we produce intervals at 95% level, the coverage is around this value only if we have at least 70% of the sample for parties PRD and PVEM, whereas for PRI we need 80% of the sample and for PAN we need 100% of the sample. Therefore, it is clear that, for incomplete samples, the credibility level has to be adjusted, if we want to achieve a 95% coverage. The credibility correction required is different according to the particular party. If we have 50% of the sample, for none of the parties we achieve a 95% coverage, even when taking a 99% credibility level. For 60% of the sample, we achieve the desired

Table 2 Coverage of credible intervals at different nominal levels by considering incomplete samples. All numbers reported are in percentages

% Sample	Nominal credible level																			
	99	98	97	96	95	99	98	97	96	95	99	98	97	96	95	99	98	97	96	95
	PAN					PRD					PRI					PVEM				
50	46	37	26	20	15	87	76	65	59	57	87	80	74	63	63	93	89	87	83	83
60	57	39	35	28	28	96	93	89	85	83	96	91	89	83	70	98	96	87	85	85
70	65	54	43	39	30	99	99	97	97	96	96	93	89	85	83	98	98	98	96	96
80	96	91	78	63	57	99	99	98	97	96	99	99	98	96	93	99	98	98	96	96
90	98	98	91	89	84	99	99	99	98	97	99	99	99	97	98	99	99	99	97	98
100	99	99	98	97	96	100	100	99	98	97	100	100	99	98	98	100	100	99	98	98

Table 3 Credibility level adjustment according to percentage of sample received

Percentage	[0, 60)	[60, 70)	[70, 80)	[80, 90)	[90, 100]
Credibility	99	98	97	96	95

coverage, in three of the four parties, by taking a 99% credibility level. For 80% of the sample we achieve the desired coverage, in three of the four parties, by taking a 96% credibility level. According to these behaviors, we suggest starting producing intervals at a 99% credibility level and gradually reducing the credibility level down to 95% as more data becomes available according to the numbers reported in Table 3.

5 Election Day

The electoral authority in Mexico, National Electorate Institute, has the responsibility of communicating precise results the night of election day. Counting all votes of all polling stations on the same day of election is practically unfeasible and therefore raises the need for robust mechanisms such as the quick count committee to estimate the corresponding trends. In particular, in the most recent 2021 election, the quick count committee had to estimate trends for 15 state governorships, the composition of the COD as well as citizen participation (i.e., the percentage of people who went out to vote).

As outlined in Sect. 3, a stratified random sample of 6,000 polling stations was selected at the national level with 20 polls in each of the $N = 300$ strata (districts). Polling stations close at 18:00 h local time, and at that moment, scrutiny of the votes starts. Once the votes have been counted, if the polling station is in the sample, a specific supervisor registers the voting results and communicates them to the corresponding district board where an operator captures the information inside a centralized database.

Each 5 min, a new (cumulative) snapshot of the database is available for the committee to produce the corresponding estimates; as expected, this snapshot doesn't include the complete sample, so incomplete sample estimation procedures, like the one described in Sect. 4, have to be implemented. The arrival of information during election day (June 6th, 2021), in terms of sample size and strata after 18:30 h, is depicted in Fig. 4. The speed at which the strata accumulate is faster than the speed of the total sample. It can be seen that at 21:30 h, we still have some strata without any available information.

Every 5 min, we produced Bayesian estimates using the model described in Sect. 2 and the incomplete sample estimation procedure of Sect. 4. Posterior estimates were obtained using a sample of size 10,000 from the posterior distribution. Sampling was required in order to incorporate the truncation in the posterior distribution (1).

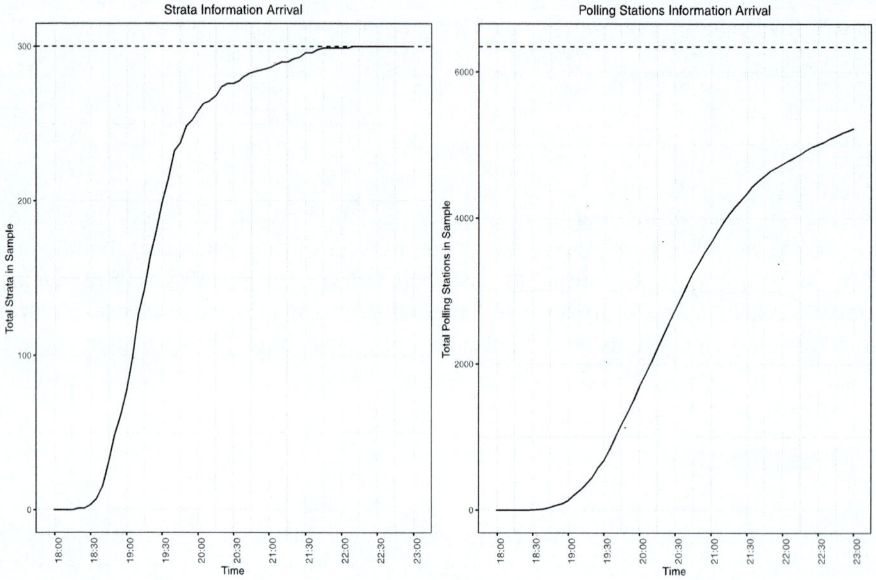

Fig. 4 Time series of cumulated sample information. Number of strata (left), number of polling stations (right)

Figure 5 illustrates the time series of interval estimates produced with each snapshot of the information available. The real values of the total number of seats per political party constituting the COD are marked with a horizontal dashed line. For most of the parties, at around 19:30 h, posterior intervals contain the true number of seats, however, there are two parties, MORENA (first row, fourth column) and PRD (second row, third column) that had to wait until 21:00 h so that our estimates contain the true values. In some other elections, final estimates are produced with missing strata; this was not the case for this election because at 22:35 h, when the final results were produced, we had information in all 300 strata, but only 79.4% of the total number of polling stations in the sample, so still an incomplete sample.

To place our modeling proposal in context, we compare it with the correction procedure proposed by [6], in the same volume, to estimate the conformation of the COD in the 2021 federal election. In Fig. 2, [6] presents time series interval estimates for the total of seats per political party. In terms of uncertainty quantification, his intervals present a similar width as compared to ours. However, there are some differences. For parties like MC, PRD, PRI, PT and PVEM, our model captures the true values earlier (i.e., with reduced sample). The only party where Rodriguez's model behaved better is MORENA. For the rest of the parties, both models behave similarly.

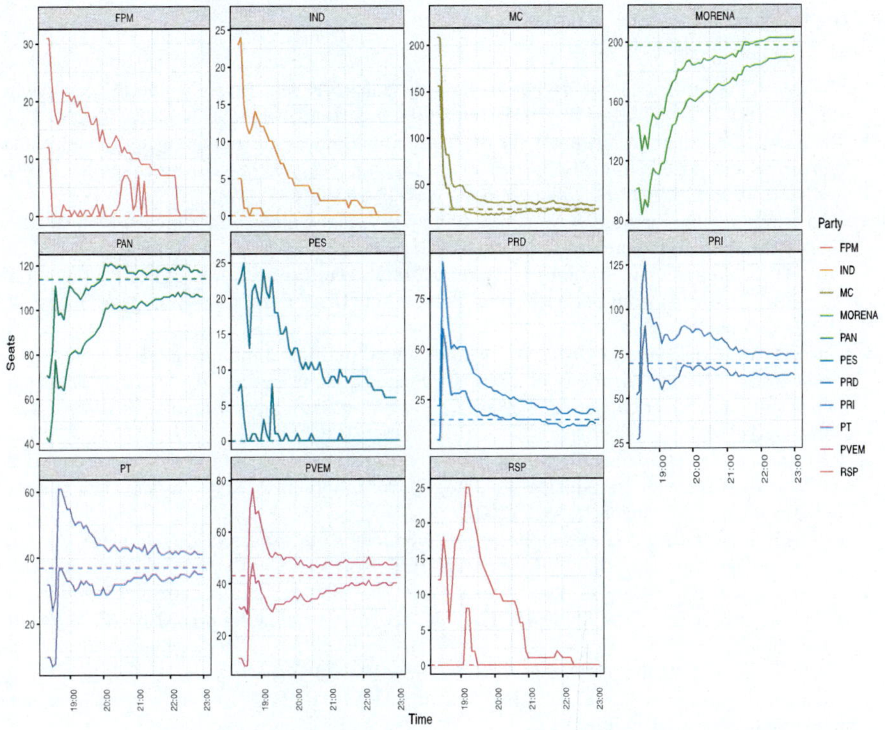

Fig. 5 Time series interval estimates of the total of seats per political party

6 Conclusions

Given the problem of an incomplete sample, a poststratification is carried out based on a series of hierarchical clusterings, which allows for a dynamic level of granularity in each update. This makes it possible to ensure that the most granular partition corresponds to the original stratification of the sample design. Hierarchical poststrata were used to impute the sufficient statistics of the missing districts. A further credible level correction according to the percentage of sample received was also included. Our procedure led to good enough estimates, despite an incomplete sample, so it is valuable to assess these problems with such techniques. However, an improved model could be developed, including the search of better clustering mechanisms through Bayesian models.

Throughout this work, it was important to assume that the volume of the nominal list as well as the electoral preferences do not change much between federal elections. The characteristics used for the clustering may change; in such a case, electoral preference should be considered along with other economic and spatial characteristics.

References

1. Glasgow, G.: Stratified sampling types. In: Kempf-Leonard, K. (ed.) Encyclopedia of Social Measurement, pp. 683–688. Elsevier (2005). https://doi.org/10.1016/B0-12-369398-5/00066-9
2. Kennedy, L., Gelman, A.: Know your population and know your model: using model-based regression and poststratification to generalize findings beyond the observed sample (2019). arXiv no. 1906.11323. https://doi.org/10.1016/0022-2836(81)90087-5
3. Stan Development Team: Stan Modeling Language Users Guide and Reference Manual 2.28 (2020). https://mc-stan.org
4. Park, D.K., Gelman, A., Bafumi, J.: Bayesian multilevel estimation with poststratification: state-level estimates from national polls. Polit. Anal. **12**, 375–85 (2004). https://doi.org/10.1093/pan/mph024
5. Anzarut, M., Gonzalez, L., Ortiz, M.: A heavy-tailed multilevel mixture model for the quick count in the Mexican elections of 2018. In: Selected Contributions on Statistics and Data Science in Latin America, pp. 1–13. Springer PROMS (2019). https://doi.org/10.1007/978-3-030-31551-1_1
6. Rodríguez, C.E.: Estimating the composition of the chamber of deputies in the quick count for the 2021 federal election in Mexico. In: Selected Contributions on Statistics and Data Science in Latin America. Springer PROMS (2022)
7. LEGIPE: Ley general de instituciones y procedimientos electorales (2017). https://www.ine.mx/wp-content/uploads/2020/07/Despen-LEGIPE-NormaINE.pdf
8. Mendoza, M., Nieto-Barajas, L.E.: Quick counts in the Mexican presidential elections: a Bayesian approach. Elect. Stud. **43**, 124–132 (2016). http://dx.doi.org/10.1016/j.electstud.2016.06.007
9. Cochran, W.G.: Sampling Techniques. Wiley, New York (1977)
10. National Electoral Institute: Proceso electoral 2021. INE (2021). https://www.ine.mx/numeralia-proceso-electoral-2021/
11. Everitt, B.S., Landau, S., Leese, M., Stahl, D.: Cluster Analysis. Wiley Series in Probability and Statistics, 5th ed., p. 88 (2010). https://doi.org/10.1002/9780470977811
12. Galili, T.: dendextend: an R package for visualizing, adjusting and comparing trees of hierarchical clustering. Bioinformatics **31**, 3718–3720 (2015). https://doi.org/10.1093/bioinformatics/btv428
13. R Core Team: R: A language and environment for statistical computing. R Foundation for Statistical Computing, Vienna, Austria (2021) . ISBN 3-900051-07-0. http://www.R-project.org/
14. Wickham, H., Averick, M., Bryan, J., Chang, W., McGowan, L.D., François, R., Grolemund, G., Hayes, A., Henry, L., Hester, J., Kuhn, M., Pedersen, T.L., Miller, E., Bache, S.M., Müller, K., Ooms, J., Robinson, D., Seidel, D.P., Spinu, V., Takahashi, K., Vaughan, D., Wilke, C., Woo, K., Yutani, H.: Welcome to the tidyverse. J. Open Source Softw. **4** (43), 1686 (2019). https://doi.org/10.21105/joss.01686
15. Lawless, J.F.: Statistical Models and Methods for Lifetime Data. Wiley, New Jersey (2003)

Maximum Likelihood Estimation for a Markov-Modulated Jump-Diffusion Model

Bor Reynoso, F. Baltazar-Larios, and Laura Eslava

1 Introduction

Modeling the stock market is a central topic of mathematical finance due to the importance of predicting values to help investors identify investments with better returns. It is generally assumed that the stock prices may be represented by a continuous-time stochastic process $\{S_t\}_{t\geq 0}$ and that we may observe its values at discrete times. From these historical data, we may estimate uncertainties and predict stock market trends to help investors when making decisions.

One of the earliest models for $\{S_t\}_{t\geq 0}$ is the Geometric Brownian Motion (GBM) introduced by Black and Scholes in 1973 [4]. This process is defined by its initial value, drift, and volatility; see Eq. (1) for a detailed definition.

The GBM is central to describing the behavior of a stock price in option pricing models. However, since the works in [5, 14, 15], it is recognized that the market dynamics cannot be modeled by GBM with constant parameters of drift and volatility. One of its main drawbacks is that large and abrupt changes in the prices may not be accounted for without assuming that the process has an artificially large volatility. The source of these abrupt changes may be macroeconomic events such as changes in the rate of economic growth, gross domestic product, unemployment rates, and inflation.

In 1976, Merton [17] proposed a JDM for asset pricing to incorporate rare and abrupt changes in the values of $\{S_t\}_{t\geq 0}$. The JDM adds, to the GBM, a compound Poisson process where jumps are independent and identically distributed; see Eq. (2) for a detailed definition. In Merton's model, the logarithm of a jump is normally distributed [17], which renders some computations relatively easier. On the other

B. Reynoso (✉) · F. Baltazar-Larios
FC, UNAM, Mexico City, Mexico
e-mail: borreynoso@ciencias.unam.mx

L. Eslava
IIMAS, UNAM, Mexico City, Mexico

I. Antoniano-Villalobos et al. (eds.), *Interdisciplinary Statistics in Mexico*,
Springer Proceedings in Mathematics & Statistics 397,
https://doi.org/10.1007/978-3-031-12778-6_11

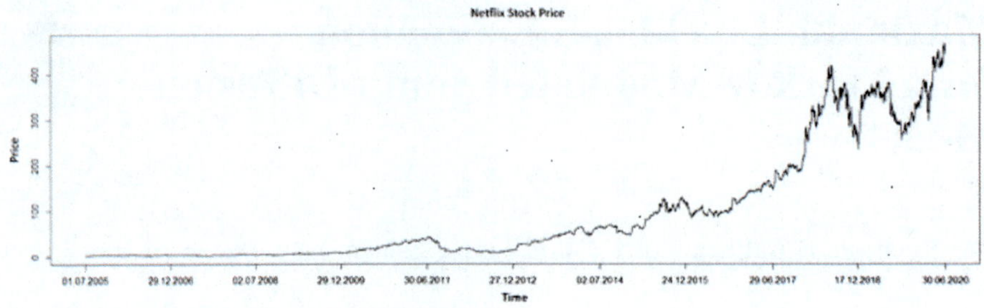

(a) Netflix Stock Price (USD) from July 1st 2005 to June 30 2020.

(b) Amazon Stock Price (USD) from July 1st 2005 to June 30 2020.

Fig. 1 Amazon and Netflix stock prices, 15 year comparison

hand, Kou [13] modifies Merton's JDM by replacing the normal distribution with the Laplace distribution; this becomes computationally more stable since the latter distribution is determined by one parameter only. Nevertheless, the continuous term of these models has constant drift and volatility which does not completely reflect the behavior of real data; see, e.g., Fig. 1.

Later models introduce random drift and random volatility parameters. For example, in [10, 12], the authors introduced the MMJDM which seeks to incorporate more variables that may influence the values of the stock prices; see Fig. 2 for an example. In the MMJDM, there is an underlying MJP that modulates both the drift and volatility of $\{S_t\}_{t\geq 0}$. The heuristics for this mechanism is that, abrupt jumps in the prices correspond to changes in the macroeconomic state of the market, therefore, not only a discontinuity occurs but also the process enters a distinct phase that determines distinct drift and volatility. We assume that the state space of the MJP represents the different environments of the financial market and so it influences the drift and volatility of the diffusion. MMJDMs for asset dynamics have been investigated in many works including [6, 7, 9].

In this paper, we are concerned with the estimation of parameters of a MMJDM. Inference, when there are two states in the underlying MJP, was presented in [8]. Likelihood-based estimation (including Bayesian) for MMJDM in risk theory has been investigated by [1, 2]. Our main contribution is Sect. 3, where we present a

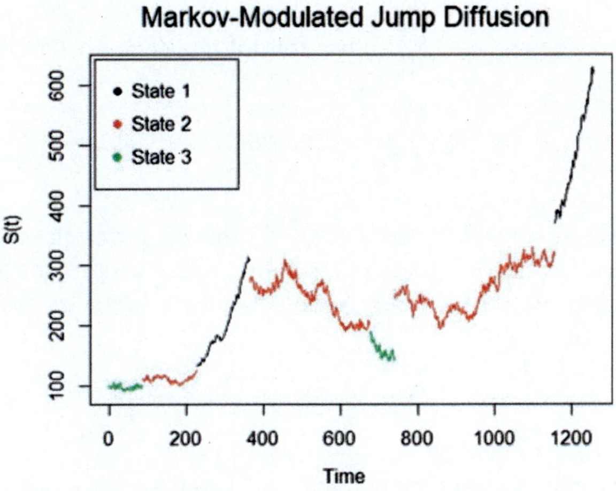

Fig. 2 Path sample of a MMJDM; see Sect. 4 for details on the parameters

methodology for the inference of the MMJDM for the general case with $m \in \mathbb{N}$ states and Laplace distribution for the logarithm of each jump; see Sect. 2 for a detailed definition. We deal with maximum likelihood estimation in the situation where we observe the diffusion process (at discrete times) and no direct observations of the MJP are given. Therefore, the data can be viewed as incomplete observations from a model with a tractable likelihood function where the full data set consists of a continuous time record of both the diffusion process and the MJP. We therefore use the EM-algorithm (see, e.g., [16, 19]) to tackle the difficulties that arise from using incomplete information in finding the MLEs. Another natural methodology used to estimate the parameters is through Markov chain Monte Carlo (MCMC) algorithms (see, e.g., [1]). Although the MCMC method allows us to obtain more information on the parameters from their distribution, in general, EM algorithms have two advantages: the speed of convergence and precision of the parameters are better and the estimation does not depend on the choice of the prior distribution.

In the next section, we define the MMJDM. The methodology to estimate the parameters of the model is detailed in Sect. 3. Sections 4 and 5 contain the simulation study and an analysis of real data from Amazon and Netflix, respectively; we close with several conclusions in Sect. 6.

2 Markov-Modulated Jump-Diffusion Model

Before introducing the MMJDM, we present the definitions of both the GBM and the JDM. The GBM is the stochastic process $\{S_t\}_{t\geq 0}$ solution of the stochastic differential equation (SDE)

$$dS_t = \mu S_t dt + \sigma S_t dB_t, \tag{1}$$

where $S_0 > 0$, $\mu \in \mathbb{R}$, $\sigma > 0$, and $\{B_t\}_{t\geq 0}$ is a standard Brownian motion.

The process $\{S_t\}_{t\geq 0}$ is a JDM if it is given by the SDE

$$dS_t = \mu S_t dt + \sigma S_t dB_t + d\sum_{k=1}^{N_t} \Delta S_k, \tag{2}$$

where, in addition to the initial value S_0, parameters μ, σ and $\{B_t\}_{t\geq 0}$ we also have a compound Poisson process $\sum_{k=1}^{N_t} \Delta S_k$ with intensity $\lambda > 0$ and i.i.d. jump sizes ΔS_k. We recall, as a warm-up, that the solution for the JDM (2) is

$$S_t = S_0 \exp\left(\left(\mu - \frac{1}{2}\sigma^2\right) t + \sigma B_t + \sum_{k=1}^{N_t} \Delta S_k\right).$$

For the MMJDM, we assume that, for some positive integer m, the financial market undergoes a macroeconomic environment that evolves among m distinct states. The financial environment is then given by a MJP $\{X_t\}_{t\geq 0}$ with state space $E := \{1, \ldots, m\}$ and intensity matrix $Q := \{q_{ij}\}_{i,j\in E}$. Let $\{\tau_k\}_{k\geq 1}$ and $\{C_t\}_{t\geq 0}$ be, respectively, the jump times and the counting process that records the number of jumps of $\{X_t\}_{t\geq 0}$.

Each of the jumps of the MJP represents the occurrence of abrupt events that cause changes in the environment and we assume that state $i \in E$ represents the i-th most favorable and stable environment, from good growth and confidence to crisis in the business sector.

For the stock prices $\{S_t\}_{t\geq 0}$, we assume to have an initial state $S_0 > 0$ and vectors $\boldsymbol{\mu} := (\mu_1, \ldots, \mu_m)$ and $\boldsymbol{\sigma} := (\sigma_1, \ldots, \sigma_m)$ corresponding to the drift and volatility of the stock prices at each state $i \in E$. In addition, let $\{J_k\}_{k\in \mathbb{N}}$ be a succession of independent random variables with Laplace distribution $(0, \eta)$; these variables will determine the jumps of the stock prices. The MMJDM associated with the stock prices is the solution to the SDE given by

$$dS_t = \mu_{X_t} S_t dt + \sigma_{X_t} S_t dB_t + d\sum_{k=1}^{C_t} \Delta S_k \tag{3}$$

where $\Delta S_k := \Big(\exp(J_k) - 1\Big) S_{\tau_k^-}$, and $\{B_t\}_{t\geq 0}$ is independent of $\{X_t\}_{t\geq 0}$. Note that the parameters of the MMDJM are given by $\boldsymbol{\theta} := (Q, \boldsymbol{\mu}, \boldsymbol{\sigma}, \eta)$.

3 Methodology for the Maximum Likelihood Estimators

We deal with the situation where we do not have a full observation of the MMJDM. The data are observations S_{t_i} of the stock prices at discrete times $0 = t_0 < t_1 \ldots <$

$t_M =: T$, where $\Delta := t_i - t_{i-1}$ for $i \in \{1, \ldots, M\}$. Our aim is to find the MLE of $\boldsymbol{\theta}$ when we only observe $\boldsymbol{S} := \{S_{t_i}\}_{i=0}^{M}$; we consider this data as an incomplete observation of a full data set consisting of both sample paths $\boldsymbol{S}^c := \{S_t\}_{t\in[0,T]}$ and $\boldsymbol{X}^c := \{X_t\}_{t\in[0,T]}$.

First, we find the likelihood function for the full data set and then estimate conditional expectation of this full log-likelihood function given the observations $\boldsymbol{S}$. Assume that we count with $(\boldsymbol{X}^c, \boldsymbol{S}^c)$; that is, a continuous observation of the MJP and historical data of the stock prices. Then the likelihood function is given by

$$L_T^c(\boldsymbol{\theta}) := \mathbb{P}(\boldsymbol{S}^c, \boldsymbol{X}^c; \boldsymbol{\theta}) = \mathbb{P}(\boldsymbol{S}^c | \boldsymbol{X}^c; \boldsymbol{\theta})\mathbb{P}(\boldsymbol{X}^c; \boldsymbol{\theta}),$$

which may be expressed as

$$L_T^c(\boldsymbol{\theta}) = \prod_{i=1}^{m}\left[\prod_{j=1}^{K}\eta e^{-\eta\ell_j}\left(\prod_{j=1}^{N_i}\prod_{k=2}^{r_{ij}}\frac{\exp\left(-\frac{(y_{ijk}-\delta_j\Delta)^2}{2\Delta\sigma_j^2}\right)}{\sqrt{2\pi\Delta\sigma_j^2}}\left(\prod_{j\neq i}^{M}q_{ij}^{N_{ij}}e^{-q_{ij}R_i}\right)\right)\right], \tag{4}$$

where

- N_{ij} is the number of jumps from state i to j,
- N_i is the number of visits to state i, i.e., $N_i := \sum_{j=1}^{m} N_{ij}$,
- R_i is the total time spent at state i,
- K is the number of jumps $K := \sum_{i=1}^{m} N_i$,
- r_{ij} is the number of observations of $\boldsymbol{S}$ on the j-th visit of $\boldsymbol{X}$ at state i,
- $y_{ijk} := \log(\frac{s_{ijk}}{s_{ij(k-1)}})$ where s_{ijk} is the k-th observation of stock prices on the j-th visit of $\boldsymbol{X}$ at state i,
- ℓ_j is the size of the j-th jump, and
- $\delta_j := \mu_j - \frac{1}{2}\sigma_j^2$.

In the case where the MMJDM is continuously observed, which is not a realistic assumption, finding the MLE is straightforward. For the case of incomplete data, the overall methodology is summarized in Algorithm 1; we detail each of its steps in the subsequent sections.

Algorithm 1 General algorithm when there are m macroeconomic states

1: Select a threshold to identify the jumps of the MJPv.
2: Estimate the coefficient of the jumps, i.e., compute the MLE of η.
3: Estimate the coefficients of the GBM, i.e., compute the MLE of μ and σ.
4: Estimate the intensity matrix of the MJP, i.e., compute the MLE of Q.

3.1 *Identify the Jumps of the MJP*

We will transfer the analysis of $\boldsymbol{S}$ to the log-yields $\boldsymbol{Y} := \{y_1, \ldots, y_M\}$ which are defined by $y_i = \log(\frac{S_{t_i}}{S_{t_{i-1}}})$, for $i = 1, \ldots, M$.

Observe that if there are no jumps in $(t_{i-1}, t_i]$, we may write $X_t = j$ for $t \in (t_{i-1}, t_i]$ and some $j \in E$ and so the process $\{S_t\}_{t\in(t_{i-1},t_i]}$ may be regarded as a GBM process with parameters μ_j and σ_j; see (1).

Thus, conditional on not having jumps in $(t_{i-1}, t_i]$ and using that $\{B_t\}_{t\geq 0}$ and $\{X_t\}_{t\geq 0}$ are independent, we can solve the SDE in (3) and obtain, for $t \in (t_{i-1}, t_i]$,

$$S_t = S_{t_{i-1}} \exp\left[\delta_j(t - t_{i-1}) + \sigma_j(B_t - B_{t_{i-1}})\right]. \tag{5}$$

Consequently, for $i = 1, \ldots, M$, assuming that $X_t = j \in E$ for $t \in [t_{i-1}, t_i]$ (i.e., there are no jumps in such interval), we infer that $y_i \sim N(\delta_j \Delta, \sigma_j^2 \Delta)$; that is,

$$y_i = \delta_j \Delta + \sigma_j(B_{t_i} - B_{t_{i-1}}). \tag{6}$$

Heuristically, the logarithm of each jump J_k should be significantly larger than the fluctuations of the diffusion over intervals of length Δ, at any of the MJP states. Namely, we assume that η^{-1} is significantly larger than $\max\{|\delta_j| : j \in E\}$. Consequently, we choose a suitable constant $U > 0$ and define $\boldsymbol{Y}_J := \{y_i \in Y : |y_i| > U\}$; we determine that $\boldsymbol{S}^c$ has (at least) a jump in the time interval $(t_{i-1}, t_i]$ whenever $|y_i| > U$.

3.2 *Estimate the Distribution of the Jumps*

Write $\boldsymbol{Y}_J := \{y_{i_1}, \ldots, y_{i_K}\}$. By assuming that there is at most one jump in each of the intervals $(t_i - t_{i-1}]$, we may associate y_{i_k} with the k-th jump of the process; in fact, we heuristically argue below that $y_{i_k} \approx J_k$. Since $|J_k| \sim Exponential(\eta)$, we have that the MLE of η is given by

$$\hat{\eta} := \frac{K}{\sum_{k=1}^{K} |y_{i_k}|}. \tag{7}$$

Indeed, assume that the k-th jump is contained in $(t_{i-1}, t_i]$. We may assume that $X_{t_{i-1}} = j$, $X_{t_i} = j'$ for some distinct $j, j' \in E$. Hence, $y_i \in \boldsymbol{Y}_J$ is the sum of J_k and two perturbations; that is

$$y_i = J_k + \Delta P_{i-1,j,k} + \Delta P_{i,j',k},$$

where $\Delta P_{\ell,m,k} \sim N\left(\delta_m|t_\ell - \tau_k|, \sigma_m^2|t_\ell - \tau_k|\right)$. Note that J_k is independent of both $\Delta P_{i-1,j,k}$ and $\Delta P_{i,j',k}$. The variability of the distribution of J_k is assumed to be significantly larger than the fluctuations of the continuous part of the process in an interval of length Δ. We have that

$$\left|\frac{\Delta P_{n,1} + \Delta P_{n,2}}{J_n}\right| \approx 0$$

which implies that the perturbations may be ignored, equivalently, $y_i \approx J_k$.

3.3 Estimate the Coefficients of the GBM

Let us analyze now the yields that correspond to intervals with no jumps. If $y_i \in \boldsymbol{Y} \setminus \boldsymbol{Y}_J$ then $X_t = j$ for some $j \in E$ and $t \in (t_{i-1}, t_i]$. Therefore, (6) implies that $y_i \sim N(\delta_j \Delta, \sigma_j^2 \Delta)$. Hence, we may regard y_i as an observation of a mixture of normal distributions, and so we estimate the parameters $\boldsymbol{\mu}$ and $\boldsymbol{\sigma}$ from the observations $\boldsymbol{Y} \setminus \boldsymbol{Y}_J$.

For ease of writing, let $N := M - K$ and let $\boldsymbol{W}_N := \boldsymbol{Y} \setminus \boldsymbol{Y}_J = \{w_1, \ldots, w_N\}$ be an enumeration of the yields that preserves the original order in $\boldsymbol{Y}$. In what follows, we introduce a latent discrete variable $\boldsymbol{Z} := \{z_1, \ldots, z_N\}$ such that $z_n = j$ whenever $w_n \sim N(\delta_j, \sigma_j^2)$; in this case, we say that w_n belong to group G_j. Observe that, conditioned to having no jumps in the interval corresponding to w_n, the latent variable z_n coincides with X_t, for all t in such interval.

For classifying $\boldsymbol{W}_N$, we can view each element w_n, with $n = 1, \ldots, N$, as an observation of a random variable W with a m-component finite mixture density given by

$$f_W(w; \Psi) := \sum_{j=1}^{m} \pi_j \frac{1}{\sqrt{2\pi\sigma_j^2 \Delta}} \exp\left(-\frac{1}{2\sigma_j^2 \Delta}(w - \delta_j \Delta)^2\right), \tag{8}$$

where $\Psi := (\pi, \boldsymbol{\mu}, \boldsymbol{\sigma})$ and π_j are the mixing proportions; that is, $\sum_{j=1}^{m} \pi_j = 1$ and for $j \in E$, $\mathbb{P}(Z_n = j) =: \pi_j \geq 0$. We have that

$$\gamma_{Z_n}(j) := \mathbb{P}(Z_n = j | w_n) = \frac{\pi_j f_j(w_n; \psi_j)}{\sum_{k=1}^{m} \pi_k f_k(w_n; \psi_k)}, \tag{9}$$

where $\psi_j = (\mu_j, \sigma_j)$ and $f_j(w_n; \psi_j)$ is the probability density function of a normal distribution of parameters $(\delta_j \Delta, \sigma_j^2 \Delta)$.

We shall call $(\boldsymbol{W}_N, \boldsymbol{Z})$ the complete data set and refer to the actual observed data $\boldsymbol{W}_N$ as incomplete. We apply the EM algorithm to classify $\boldsymbol{W}_N$ and, simultaneously, approximate the MLE of both $\boldsymbol{\mu}$ and $\boldsymbol{\sigma}$. We have that the log-likelihood of the complete data (see [11]) is given by

$$\ell(\Psi | \boldsymbol{W}_N, \boldsymbol{Z}) = \ell(\Psi) = \sum_{n=1}^{N} \sum_{j=1}^{m} \log\left(\pi_j \gamma_{Z_n}(j); \psi_j)\right) + \sum_{n=1}^{N} \sum_{j=1}^{m} \log\left(f_j(w_n | Z_n = j; \psi_j)\right). \tag{10}$$

Since the increments of the Wiener process $\{B_t\}_{t \geq 0}$ are conditionally independent (on MJP), so are their corresponding log-returns and we can use the EM Algorithm 2 to estimate $\boldsymbol{\mu}$ and $\boldsymbol{\sigma}$.

Algorithm 2 EM for estimation of $\boldsymbol{\mu}$ and $\boldsymbol{\sigma}$

1: Choose an initial value $\Psi^{(0)} := (\pi_j^{(0)}, \delta_j^{(0)}, \sigma_j^{2(0)})_{j=1,\ldots m}$, tolerance $\varepsilon > 0$ and set $t = 0$.
2: **E-step**: Compute $S(\Psi; \Psi^{(t)}) := \mathbb{E}[\ell(\Psi)|\boldsymbol{W}_N; \Psi^{(t)}]$ using (10) and set

$$\gamma_{z_n}(j)^{(t+1)} = \frac{\pi_j^{(t)} f_j(w_n; \psi_j^{(t)})}{\sum_{k=1}^m \pi_k^{(t)} f_k(w_n; \psi_k^{(t)})},$$

for $n = 1, \ldots, N$ and $j \in E$.
3: **M-step**: For $j \in E$ let

$$\pi_j^{(t+1)} = \frac{\sum_{n=1}^N \gamma_{z_n}(j)^{(t+1)}}{N}, \quad \delta_j^{(t+1)} = \frac{\sum_{n=1}^N w_n \gamma_{z_n}(j)^{(t+1)}}{\sum_{n=1}^N \gamma_{z_n}(j)^{(t+1)}},$$

$$\sigma_j^{2(t+1)} = \frac{\sum_{n=1}^N (w_n - \delta_j^{(t+1)})^2 \gamma_{z_n}(j)^{(t+1)}}{\sum_{n=1}^N \gamma_{z_n}(j)^{(t+1)}}.$$

4: Compute $S(\Psi; \Psi^{(t+1)})$ using (10).
5: If $|S(\Psi; \Psi^{(t)}) - S(\Psi; \Psi^{(t+1)})| < \varepsilon$, let $\hat{\mu}_j = \delta_j^{(t+1)} + \frac{1}{2}\sigma_j^{2(t+1)}$, $\hat{\sigma}_j = \sqrt{\sigma_j^{2(t+1)}}$ and stop.
6: Let $t = t + 1$ and go to E-step.

3.4 *MLE of Q*

Finally, it remains to estimate the parameter Q of the MJP process $\{X_t\}_{t \geq 0}$. To do so, we need to estimate the time the process spends at each state $j \in E$ and the frequency of jumps from state i to state j, for distinct $i, j \in E$. First, we classify the clusters of yields between jumps, thus recovering the bulk of time spent at each state $j \in E$. The inference from the yields in which a jump is involved is more delicate and is performed in Algorithm 3.

Let us decompose $\boldsymbol{W}_N$ into $K + 1$ clusters,

$$\{w_1^0, \ldots, w_{n_0}^0\}, \{w_1^1, \ldots, w_{n_1}^1\}, \ldots, \{w_1^K, \ldots, w_{n_K}^K\},$$

where $\{w_1^k, \ldots, w_{n_k}^k\}$ are the yields between k-th and $(k+1)$-th jumps for $1 \leq k < K$ and, similarly, $\{w_1^0, \ldots, w_{n_0}^0\}$ and $\{w_1^K, \ldots, w_{n_K}^K\}$ are the yields before the first jump and after the K-th jump, respectively.

We again refer to the latent variables $\boldsymbol{Z}$ to classify each cluster in one of the m groups G_j, $j \in E$. For $j \in E$ and $0 \leq k \leq K$, we have that the probability that the block $\{w_1^k, \ldots, w_{n_k}^k\}$ comes from the j-th density is

$$\Gamma_k(j) := \mathbb{P}(z_1^k = j, \ldots, z_{n_k}^k = j | w_1^k, \ldots, w_{n_k}^k) = \prod_{\ell=1}^{n_k} \mathbb{P}(z_\ell^k = j | w_\ell^k) = \prod_{\ell=1}^{n_k} \gamma_{z_\ell^k}(j). \tag{11}$$

Thus, we partially reconstruct a path of the MJP for the time intervals $\cup_{k=0}^K [r_k^-, r_k^+]$, where r_k^- is the time of observation $w_1{}^k$ and r_k^+ is the time of observation $w_{n_k}^k$.

We classify each cluster $\{w_1^k, ..., w_{n_k}^k\}$ using $\Gamma_k(j)$ for $j \in E$. If the k-th cluster is in state j then $X_t = j$ for $t \in [r_k^-, r_k^+]$.

The values of $\{X_t\}$ are still undetermined in the missing intervals, i.e., we can only observe $\{X_t\}$ at the endpoints of interval $[r_k^+, r_{k+1}^-]$ for $k = 0, \ldots, K-1$. Hence, we use Algorithm 3 to, iteratively, simulate such missing paths and estimate the MLE of each of the entries in the intensity matrix Q. Algorithm 3 is a Stochastic EM (SEM) algorithm (see, e.g., [18]); the E-step and M-step of each of these algorithms are repeated until convergence, i.e., when the estimators are observed to arrive at the stationary distribution. An advantage of using the stochastic version of EM algorithm is that, by having the distribution of the estimator, we can calculate confidence intervals for the MLEs obtained.

Algorithm 3 SEM for the estimation of $\boldsymbol{Q}$.

1: Choose an initial value $\boldsymbol{Q}^0$ and make $\ell = 0$.
2: Using the current $\boldsymbol{Q}^\ell$, for each $0 \le k < K-1$, draw a Markov bridge between the observations $X_{r_{k-1}^+}$ and $X_{r_k^-}$ to generate a path $\boldsymbol{X}^c = \{X_t\}_{t=0}^T$.
3: **E-step** Using $\boldsymbol{X}^c$, we compute the corresponding statistics N_{ij} and R_i for all $i, j \in E$ in time interval $[0, T]$.
4: **M-step** The MLEs of q_{ij} (see [2]) are given by

$$q_{ij}^{\ell+1} = \frac{N_{ij}}{R_i}, \quad i \neq j, \quad q_{ii}^{\ell+1} = -\sum_{j \neq i} q_{ij}^{\ell+1}.$$

5: $\ell = \ell + 1$ and go to Step 2.

4 Simulation Study

In this section, we present the results of a simulation study that calibrates the method presented in Sect. 3. We test the model with $m = 3$; that is, we consider three states going from a stable economy to an unstable economy to crisis in the sector. Hence, we may consider in the test that drifts μ_i are decreasing (possibly $\mu_m < 0$) while volatilities σ_i are increasing in $i \in E$.

We consider a MMJDM $\{S_t\}_{t \ge 0}$ with the following parameters. Let

$$Q = \begin{pmatrix} -0.002645503 & 0.002314815 & 0.00033068785 \\ 0.003968254 & -0.005952381 & 0.0019841270 \\ 0.005952381 & 0.001984127 & -0.0079365079 \end{pmatrix}, \tag{12}$$

$$\boldsymbol{\mu} = (0.0059523810, 0.0011904762, -0.0009920635), \tag{13}$$

$$\boldsymbol{\sigma} = (0.009449112, 0.017817415, 0.022047928), \tag{14}$$

$$\eta = 250/33. \tag{15}$$

Table 1 MLE of η

	Value	Error
η	7.575757	
$\hat{\eta}^{1260}$	5.897666	1.678092
$\hat{\eta}^{2520}$	6.241154	1.334603
$\hat{\eta}^{3780}$	7.756247	0.1804897
$\hat{\eta}^{5040}$	7.352301	0.2234567
$\hat{\eta}^{6300}$	7.144313	0.4314449
$\hat{\eta}^{7560}$	6.994469	0.5812881
$\hat{\eta}^{8820}$	7.175147	0.4006109

Table 2 MLE of μ_j

	$j=1$	$j=2$	$j=3$	Error
μ_j	0.0059523810	0.0011904762	−0.0009920635	
$\hat{\mu_j}^{1260}$	0.0059469097	0.0027668870	0.0019743179	0.00001128
$\hat{\mu_j}^{2520}$	0.0066422130	0.0034594822	0.0025009336	0.00001782
$\hat{\mu_j}^{3780}$	0.0066102728	0.0032372128	0.0020031847	0.00001359
$\hat{\mu_j}^{5040}$	0.0063053813	0.0028402039	0.0020858060	0.00001231
$\hat{\mu_j}^{6300}$	0.0059171341	0.0031332369	0.0017120427	0.00001108
$\hat{\mu_j}^{7560}$	0.0060495312	0.0030009829	0.0013917915	0.00000897
$\hat{\mu_j}^{8820}$	0.0060614045	0.0029881755	0.0008520252	0.00000664

We generate a sample path $\boldsymbol{S} = \{S_{t_i}\}_{i=0}^{M}$ where $S_0 = 100$, $X_0 = 1$, $\Delta = 1$ and $T = 1260, 2520, 3780, 5040, 6300, 7560, 8820$. Observe that if $\Delta = 1$ represents one-day periods, then $\boldsymbol{S}$ represents, for example, the daily observations of the stock prices during a 7-year period of time if $T = 8820$. In fact, the parameters $\boldsymbol{\theta}$ may be proposed on a time scale of 1 year (252 business days) and then rescaled to (12)–(15) which then may be interpreted as daily rates of the process.

Figure 2 shows a path of MMJDM $\{S_t\}_{t\geq 0}$ with the aforementioned parameters in the time interval [0, 1260]. The MLEs of η are presented in Table 1. The MLEs of $\boldsymbol{\mu}$ and $\boldsymbol{\sigma}$ for 1000 iterations are given in Tables 2 and 3 respectively; Algorithm 2 was run with $\epsilon = 0.05$. We tried several seeds $\boldsymbol{\Psi}^{(0)}$, $\boldsymbol{Q}^{(0)}$ for Algorithms 2 and 3. In this study, no dependence was observed in the estimations except for the number of iterations, so we report the results using the seeds that best represent the model.

Finally, the MLE of the intensity matrix (after 1000 iterations) when $T = 8820$ is given in Eq. (16). Figure 3 plots the iterations of the Algorithm 1 to find the MLE of $\hat{Q}$ and 95% confidence intervals; we estimate the standard deviation of the MLEs through the Fisher information matrix (see [3]). We obtain

Table 3 MLE of σ_j

	$j = 1$	$j = 2$	$j = 3$	Error
σ_j	0.009449112	0.017817415	0.022047928	
$\hat{\sigma}_j^{1260}$	0.012769167	0.015921710	0.017703704	0.00003348
$\hat{\sigma}_j^{2520}$	0.009535770	0.015774837	0.017788880	0.00002231
$\hat{\sigma}_j^{3780}$	0.009327915	0.015878260	0.018081729	0.00001950
$\hat{\sigma}_j^{5040}$	0.009104289	0.015999326	0.017911584	0.00002053
$\hat{\sigma}_j^{6300}$	0.009154862	0.015454727	0.018728368	0.00001668
$\hat{\sigma}_j^{7560}$	0.009237009	0.014904877	0.014904877	0.00001700
$\hat{\sigma}_j^{8820}$	0.014904877	0.014492026	0.019310947	0.00001854

$$\hat{Q} = \begin{pmatrix} -0.003215818 & 0.002969198 & 0.0002466196 \\ 0.003882034 & -0.005754555 & 0.0018725209 \\ 0.006037886 & 0.002097409 & -0.0081352957 \end{pmatrix}. \tag{16}$$

5 Real Data

In this section, we fit the MMJDM for two sets of stock price real data using the method presented in Sect. 3. Each data set corresponds to Amazon and Netflix daily stock prices from July 1, 2005, to June 30, 2020; see Fig. 1. Thus, it is natural to consider a day as the unit of time; that is $\Delta = 1$. This historical data presents abrupt changes and these do not occur homogeneously in time and so we may use the MMJDM model to fit the data; based on the graphical evidence, we assume that the underlying MJP has three states.[1]

Following our method, the first step is to identify the jumps. We can observe in the final prices of Netflix's stock that the price suffers a change in its behavior after March 20, 2020. This can be explained because in March many countries introduced lockdown measures due to the COVID-19 pandemic. In order to catch this change in this behavior, the threshold U for the jumps was set to $\pm\sqrt{0.01174792}$ for the squared log-returns. Similarly, the tolerance level for Amazon was set to $\pm\sqrt{0.005068828}$. As we do not consider jumps in consecutive days, we have 39 jumps for Netflix and 41 jumps for Amazon. These were used to obtain the jump parameter estimations $\hat{\eta} = 6.359798$ for Netflix and $\hat{\eta} = 9.119797$ for Amazon.

Algorithm 2 was run considering 1000 iterations; we use $\epsilon = 0.05$ and initial values as in (13) and (14). The MLEs obtained for the drift and volatility parameters for both Netflix and Amazon are shown in Table 4.

[1] Estimating the number of states is a classification problem is more complex and beyond the scope of this paper.

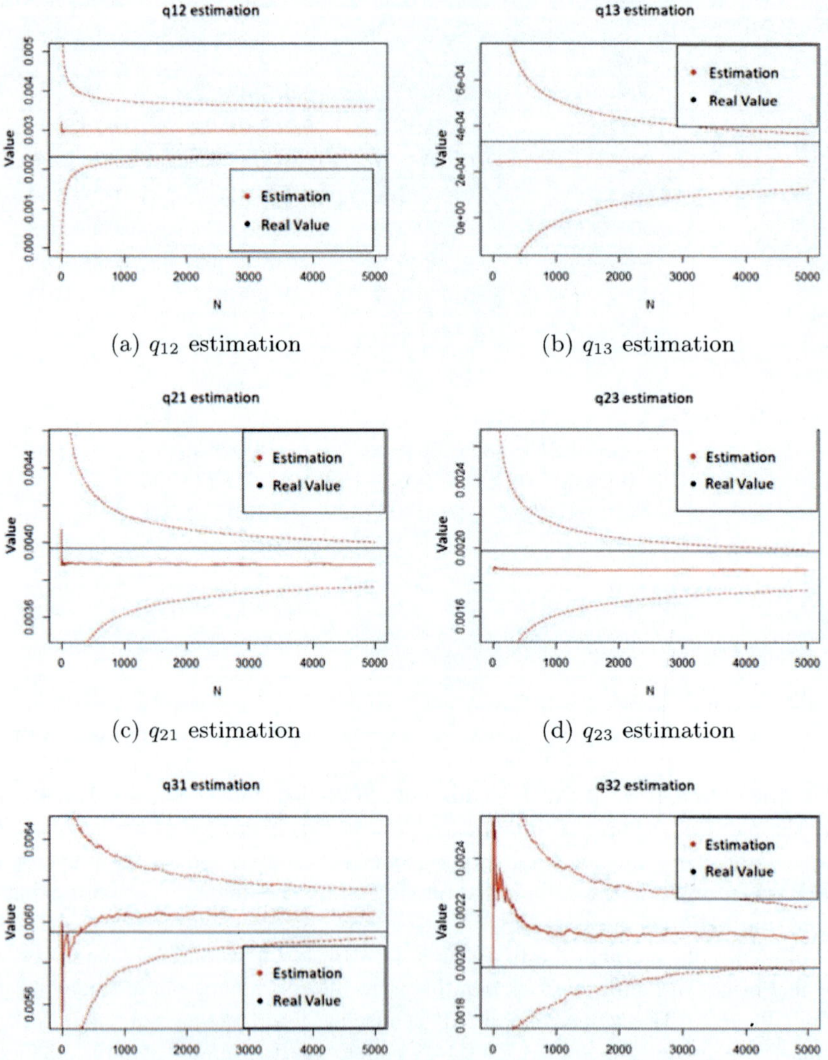

(a) q_{12} estimation (b) q_{13} estimation

(c) q_{21} estimation (d) q_{23} estimation

(e) q_{31} estimation (f) q_{32} estimation

Fig. 3 Estimation of Q for the simulation study with $T = 8820$ and 95% confidence intervals (dotted line)

Table 4 MLE of the weights, drifts and volatilities for Netflix and Amazon obtained with Algorithm 2 with tolerance level $\epsilon = 0.05$

(a) Netflix

j	$\hat{\pi}_j$	$\hat{\mu}_j$	$\hat{\sigma}_j$
1	0.44797175	−0.00026218670	0.01402849
2	0.53779080	0.0031193879	0.03213458
3	0.01423745	−0.00634970790	0.09737263

(b) Amazon

j	$\hat{\pi}_j$	$\hat{\mu}_j$	$\hat{\sigma}_j$
1	0.45787160	0.0009578106	0.01093079
2	0.48193159	0.0012029475	0.02213378
3	0.06019681	0.0041515128	0.04834662

Once the drifts and volatilities are estimated, we classified each cluster using (11) and partially constructed the path of the MJP for each stock. The initial value for both Q_{NFLX} and Q_{AMZN} estimation was (12) and we set the EM algorithm to 1000 initial iterations. The MLE for Q_{NFLX} and Q_{AMZN} is shown in equations (17) and (18), respectively. Finally, the graphs of the estimations of Q_{NFLX} and Q_{AMZN} are shown in Figs. 4 and 5, respectively.

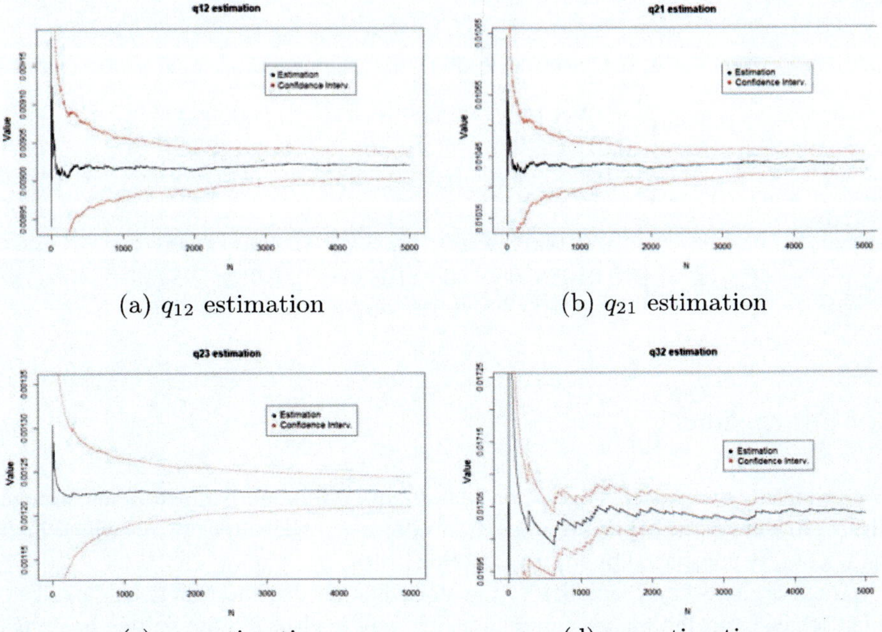

(a) q_{12} estimation (b) q_{21} estimation

(c) q_{23} estimation (d) q_{32} estimation

Fig. 4 Estimation of Q for Netflix stock

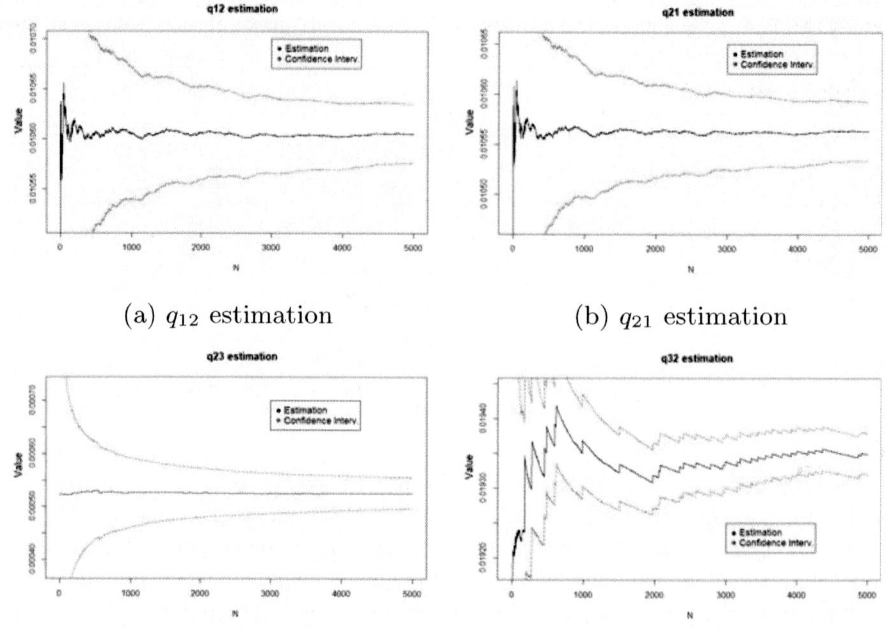

(a) q_{12} estimation

(b) q_{21} estimation

(c) q_{23} estimation

(d) q_{32} estimation

Fig. 5 Estimation of Q for Amazon stock

$$\hat{Q}_{NFLX} = \begin{pmatrix} -0.009019071 & 0.009019071 & 0.000000000 \\ 0.010438460 & -0.011664262 & 0.001225801 \\ 0.000000000 & 0.017037859 & -0.017037859 \end{pmatrix}, \tag{17}$$

$$\hat{Q}_{AMZN} = \begin{pmatrix} -0.01060526 & 0.01060526 & 0.0000000000 \\ 0.01056306 & -0.01108866 & 0.0005255957 \\ 0.00000000 & 0.01934920 & -0.0193491966 \end{pmatrix}. \tag{18}$$

6 Conclusions

We propose a methodology based on Maximum Likelihood Estimators to estimate the parameters of the MMJDM process; this process models stock prices whose drift and volatility are shaped by macroeconomic states.

There are two key properties of the stock market that the MMJDM describes. First, changes in the macroeconomic states may lead to discontinuities or jumps in the prices (disguised in the sample paths as abrupt fluctuations). Second, when these changes occur, they also modify the overall diffusion properties of the prices.

Although previous Jump-Diffusion Models incorporate the eventual jumps in the process, the MMJDM also takes into account the changes in drift and volatility that fortuitous events or news in the sector may entail.

Among the challenges faced in this work are identifying the jumps, classifying the clusters of the log yields, and optimizing the EM algorithms. We highlight that, as the simulation study in Sect. 3 shows, the methodology is robust since we are able to recover the parameters of the synthetic MMJDM.

The study with the real data on both Amazon's and Netflix's prices qualitatively reflects the same macroeconomic fluctuations. It is natural to conjecture that if the states are ordered according to the degree of stability in the market, then the transitions between the stability and crisis will be sequential (e.g., before a crisis hit, there is a period of increased volatility). It is remarkable to note that, although this heuristic is not built into the model, there is no transition from the stable to the crisis state (that is, $\hat{q}_{13} = \hat{q}_{31} = 0$).

Therefore, we consider that the MMJDM more accurately represents the nature of fluctuations in the stock market. Moreover, the methodology proposed in this work to estimate the parameters of the process allows us to make inferences about the financial environments that favor a stock, which is a useful tool in stock value speculation.

As future work, one may generalize this method by replacing the JMP with an underlying process where the length of stay in each stage follows a Phase Type distribution. One may also fit the distribution of the jump sizes according to the observed data. Finally, one may also consider methodologies that test for a suitable number of macroeconomic states since, at the moment, we assume the number of states m is given.

Acknowledgements We would like to thank the anonymous referee for their thorough comments. This work has been supported by PAPIIT TA100820.

References

1. Baltazar-Larios, F., Esparza, L.J.R.: Bayesian estimation for the Markov-modulated diffusion risk model. In: Antoniano-Villalobos, I., Mena, R., Mendoza, M., Naranjo, L., Nieto-Barajas, L. (eds) Selected Contributions on Statistics and Data Science in Latin America. FNE 2018. Springer Proceedings in Mathematics & Statistics, vol. 301. Springer, Cham (2019). https://doi.org/10.1007/978-3-030-31551-1_2
2. Baltazar-Larios, F., Esparza, L.J.R.: Statistical inference for partially observed Markov-modulated diffusion risk model. In: Methodology and Computing in Applied Probability. Version on line (2022). https://doi.org/10.1007/s11009-022-09932-7
3. Bladt, M., Esparza, L.J.R., Friis, B.F.: Fisher information and statistical inference for phase-type distributions. J. Appl. Probab. **48**(A), 277–293 (2011). https://doi.org/10.1239/jap/1318940471
4. Black, F., Scholes, M.: The pricing of options and corporate liabilities. J. Polit. Econ. **81**(3), 637–654 (1973). https://doi.org/10.1086/260062
5. Clark, P.K.: A subordinated stochastic process model with finite variance for speculative prices. Econometrica **41**(1), 135–155 (1973). https://doi.org/10.2307/1913889

6. Costabile, M., Leccadito, A., Massabó, I., Russo, E.: Option pricing under regime-switching jump-diffusion models. J. Comput. Appl. Math. **256**, 152–167 (2014). https://doi.org/10.1016/j.cam.2013.07.046
7. Das, M.K., Goswami, A., Rana, N.: Risk sensitive portfolio optimization in a jump diffusion model with regimes. SIAM J. Control Optim. **56**(2), 1550–1576 (2018). https://doi.org/10.1137/17M1121809
8. Das, M.K., Goswami, A.N., Rajani, S.: Inference of binary regime models with jump discontinuities (2021). Preprint arXiv:1910.10606
9. Elliott, R.J., Siu, T.K., Chan, L., Lau, J.W.: Pricing options under a generalized Markov-modulated jump-diffusion model. Stoch. Anal. Appl. **25**(4), 821–843 (2007). https://doi.org/10.1080/07362990701420118
10. Guo, X.: Information and option pricings. Quant. Finance **1**, 38–44 (2001). https://doi.org/10.1080/713665550
11. Guzman Solis, C.: . Algoritmo EM: Implementación y Aplicaciones. Facultad de Ciencias UNAM (2018)
12. Jobert, A., Rogers, L.C.G.: Option pricing with Markov-modulated dynamics. SIAM J. Control. Optim. **44**(6), 2063–2078 (2006). https://doi.org/10.1137/050623279
13. Kou, S.G., Wang, H.: Option pricing under a double exponential jump diffusion model. Manag. Sci. **50**(9), 1178–1192 (2004). https://doi.org/10.1287/mnsc.1030.0163
14. Mandelbrot, B.: The variation of certain speculative prices. J. Bus. **36**(4), 394–419 (1963). https://doi.org/10.1086/294632
15. Mandelbrot, B., Taylor, H.M.: On the distribution of stock price differences. Oper. Res. **15**(6), 1057–1062 (1967). https://doi.org/10.1287/opre.15.6.1057
16. McLachlan, G.J., Krishnan, T.: The EM Algorithm and Extensions. Wiley, New York (2007). https://doi.org/10.1002/9780470191613
17. Merton, R.C.: Option pricing when underlying stock returns are discontinuous. J. Financ. Econ. **3**(1–2), 125–144 (1976). https://doi.org/10.1016/0304-405X(76)90022-2
18. Nielsen, S.F.: The Stochastic EM algorithm: estimation and asymptotic results. Bernoulli **6**(3), 457–489 (2000). 132.248.181.19
19. Tanner, M.A.: Tools for Statistical Inference: Methods for the Exploration of Posterior Distributions and Likelihood Functions. Springer-Verlag (1996)

Estimating the Composition of the Chamber of Deputies in the Quick Count for the 2021 Federal Election in Mexico

Carlos E. Rodríguez

1 Introduction

A key piece in any electoral system is the electoral referee. In Mexico, the National Electoral Institute (or INE by its Spanish acronym) is an autonomous, independent and public organization responsible for organizing federal elections. The strength of the INE is constantly questioned and accused of partiality to solving political issues. In this context of distrust, the INE has implemented three mechanisms to count the votes. First, an estimate of voting trends, known as the Quick Count. Second, a preliminary count whose progress is transmitted in real time, known as the Preliminary Results Program or PREP, and the final computations, which are known as district computations, making it clear that the first two alternatives have no official validity and are only informative. Their objective is to provide certainty, transparency and confidence in the electoral process by offering timely and robust information on the trends of the election results: downplaying the candidates' self-proclamations of victory, maintaining the population informed and, therefore, reducing pressure on the electoral referee.

1.1 The Quick Count

The Quick Count is a statistical exercise that, using information from the voting in a probabilistic sample of polling stations, seeks to estimate the voting trends in a given election and communicate them to the population on the evening of the same day of

C. E. Rodríguez (✉)
Department of Probability and Statistics at the National Autonomous University of Mexico, Mexico City, Mexico
e-mail: carloserwin@sigma.iimas.unam.mx
URL: http://sigma.iimas.unam.mx/carloserwin/

I. Antoniano-Villalobos et al. (eds.), *Interdisciplinary Statistics in Mexico*,
Springer Proceedings in Mathematics & Statistics 397,
https://doi.org/10.1007/978-3-031-12778-6_12

the election. Estimates are always made in the form of an interval and their objective is to fill the information gap that occurs when polls are closed.

The Quick Count reaches its climax on the day of the election, while the sample of polling places is selected a couple of days before. Voter Trainers and Assistants (CAEs by its Spanish acronym) are essential for the Quick Count. The CAEs are personnel hired by the INE who (1) recruit citizens to be polling officers, (2) train citizens who have agreed to serve as polling officers and (3) support them through any mishaps that may arise on Election Day. The CAEs also send the voting information from the polling stations selected in the sample.

The main objective of the Quick Count for the Federal election of June 6, 2021, was to estimate the conformation of the Chamber of Deputies, i.e. of the 500 deputies that make up the chamber, the number obtained by each political force. These estimates have to be made via 95% confidence intervals.

2 The Chamber of Deputies

The Legislative Power in Mexico is deposited in a General Congress, which is divided into two Chambers: the Chamber of Deputies (500 legislators) and the Chamber of Senators (128 legislators). In addition to the tasks it shares with the Senate, the Chamber of Deputies has exclusive authority: approve the public account of the previous year, annually approve the Federation Expenditure Budget (known as PEF), approve the National Development Plan, coordinate and evaluate the performance of the functions of the Superior Audit of the Federation, ratify the appointment of the Ministry of Finance and Public Credit, declare whether or not there is a place to proceed criminally against public servants who have committed a crime and designate the heads of the internal control bodies of the bodies with autonomy that exercise resources from the PEF.

To understand the importance of the number of seats that each party holds in the Chamber of Deputies, it is necessary to know the concepts of simple majority and qualified majority. Simple majority is obtained with 250 seats plus one, of the 500 federal deputies. With this type of majority it is possible to approve the PEF, appointments in the Commissions, remuneration of public servants, fiscal proposals, as well as reject projects submitted to the plenary session of the chambers. Meanwhile, the qualified majority is achieved when 334 votes in favor are obtained, which allows the approval of structural reforms or bills without having to seek consensus with other political forces.

2.1 Conformation

Appendix A presents a detailed description of the considerations and procedures for transforming the votes received at the polls into seats in the Deputy Chamber. In this section, only the basic notation and concepts are defined.

Throughout this work, it is assumed that at the national level there are k political parties with valid registration and r independent candidates. Then, letting political forces $= \{P_1, \ldots, P_k, I_1, \ldots, I_r\}$, the objective is to estimate the composition of the Chamber of Deputies, in other words to estimate the vector

$$\text{CONF} = (NP_1, \ldots, NP_k, NI),$$

where $NP_j \geq 0$ is the number of deputies/seats in the Chamber of Deputies obtained by the political party P_j and $NI \geq 0$ is the total number of seats obtained by the r independent candidates.

Five hundred deputies integrate the Chamber of Deputies (article 52 of the [2]): 300 elected according to the voting principle of Relative Majority (RM), plus 200 deputies who will be elected according to the principle of Proportional Representation (PR). Independent candidates can only access the Chamber of Deputies via the RM principle, while the political parties can obtain deputies via both principles. Then, the number of seats in the Chamber of Deputies of the political party P_j is given by $NP_j = M_j + R_j$, for $j = 1, \ldots, k$, where M_j and R_j are the number of deputations by the principles of RM and PR, respectively. Then, we can define vectors considering the number of deputies for each political force by each principle

$$RM = (M_1, \ldots, M_k, NI) \text{ and } PR = (R_1, \ldots, R_k, 0).$$

Thus, the following equalities must be met:

1. The sum of deputies by the RM principle is given by $\sum_{j=1}^{k} M_j + NI = 300$.
2. The sum of deputies by the PR principle is given by $\sum_{j=1}^{k} R_j = 200$.
3. The sum of deputies for both principles is given by

$$\sum_{j=1}^{k}(M_j + R_j) + NI = \sum_{j=1}^{k} NP_j + NI = 500.$$

The foregoing considers that according to article 54, numeral IV of the [2], no party may have more than 300 deputies for both principles.

We stress that this procedure is not trivial, and it is essential to carefully review several articles of both the Constitution of the United Mexican States ([2]) and of the General Law of Electoral Institutions and Procedures ([6]), as well as various documents published by the INE, for example, the agreements INE/CG804/2015[1] and INE/CG1181/2018,[2] where part of the procedure is described. There are manuals describing the method (see, for example, [5]). However, these are outdated since

[1] https://transparencia.ine.mx/obligaciones/rsc/documentos/Articulo74/Formato4d/DEPYPP/inecg8042015.pdf.

[2] https://repositoriodocumental.ine.mx/xmlui/bitstream/handle/123456789/98206/CGor201808-23-ap-5.pdf.

Mexican law is in constant change. For example, for the 2021 process the INE included what is known as effective affiliation; see the agreement INE/CG193/2021.[3] The effective affiliation is a validation that the INE makes to the coalition agreements. Here, it is verified that the coalition candidate is a member of the party. Otherwise, a correction is made and the seat is counted to the party in which the coalition candidate is active. This information is essential to obtain the number of seats per RM and due to the tight timing, the INE published these agreements just a few days before the election (see agreement INE/CG466/2021[4]).

Finally, to devise the algorithm presented in Appendix A, an effort was made to keep vectors of the same dimension throughout the algorithm. This helps to vectorize the operations in R and make the algorithm more efficient, thus allowing us to (1) perform thousands of simulations to define the sampling design and (2) use Bootstrap ideas (within a Multiple Imputation setting) to produce interval estimates for the number of seats obtained in the Chamber by each political force.

3 Estimation

Finite population sampling and simulation techniques are the main tools used to define the sampling design. While multiple imputation and bootstrap ideas are used to make inference, to define the sample design and test our methods previous to the 2021 Election, the District Computations for the Deputy Chamber of 2012, 2015 and 2018 were used:

- https://portalanterior.ine.mx/documentos/proceso_2011-2012/alterna/computos.html
- http://computos2015.ine.mx/BaseDeDatos.html?header
- https://computos2018.ine.mx/#/descargaBase

The assumption is that the variance of the estimates for the 2021 Election would behave roughly the same as the variance of the estimates of these past elections.

3.1 Sampling Design

The INE installed a total of $N = 162,809$ polling stations spread all over the country for the 2021 Election. There was an independent election over each one of the 300 federal districts in which Mexico is divided. Thus, a natural choice for strata was these federal districts. Since the result in each one of the federal districts is equally

[3] https://repositoriodocumental.ine.mx/xmlui/bitstream/handle/123456789/118249/CG2ex202103-19-ap-3.pdf.

[4] https://repositoriodocumental.ine.mx/xmlui/bitstream/handle/123456789/120002/CG2ex202105-20-ap-1-a.pdf.

important (each result determines one deputy), it was decided to set $n_h = n/300$, for $h = 1, \ldots, 300$ then only needing to define the overall sample size n, i.e. the total number of polling places to select. This helped to deal with the non-response, avoiding federal districts with too small sample sizes.

The aim of the Quick Count is to estimate the conformation of the Chamber of Deputies, i.e. $NP_1, \ldots, NP_k$, hence, we need a measure of distance between the true number of seats and its estimate $\hat{NP}_1, \ldots, \hat{NP}_k$. To assess the margin of error of our sample design, we have used two measures of distance: the average and the maximum number of seats incorrectly classified. To obtain the margin of error with these distance measures, we defined

$$\mathbb{P}\left(\frac{1}{k}\sum_{j=1}^{k}|NP_j - \hat{NP}_j| \leq \epsilon_1\right) = 0.95, \text{ and } \mathbb{P}\left(\max_{j=1,\ldots,k}|NP_j - \hat{NP}_j| \leq \epsilon_2\right) = 0.95, \tag{1}$$

where simulation techniques are used to approximate ϵ_1 and ϵ_2 for different sample sizes. A detailed description of the process is given below. Observe that this idea is the cornerstone for the bootstrap algorithm to estimate the conformation of the Chamber of Deputies:

1. Use the voting results per voting station for the Deputy Chamber Election of 2012, and via the algorithm described in Sect. 2.1, obtain the "true" conformation of the Chamber of Deputies, i.e. $NP_1, \ldots, NP_k$.
2. Set an overall sample size n.
3. Select a simple random sample of size $n_h = n/300$ voting places from district h, where $h = 1 \ldots, 300$.
4. Estimate the total number of votes obtained by the political party P_j in the federal district h via the usual ideas of stratified sampling, i.e.

$$VTP_{h,j} = (N_h)(\bar{y}_{h,j}) = \frac{N_h}{n_h}\sum_{l=1}^{n_h} y_{h,j,l},$$

   where $y_{h,j,l}$ is the number of votes obtained by candidate j in the polling place labeled as l from the random sample of federal district h. Then, the number of votes at a national level, for the political party P_j, is given by $VTP_j = \sum_{h=1}^{300} VTP_{h,j}$. **Observation**: this step can only be done after considering the coalition agreements on each federal district; see Sect. A.1.
5. Estimate the total number of votes for independent, non-registered candidates and null votes via the exact same ideas as before to obtain VTI, $VTNR$ and VTN, respectively. Then, estimate the total number of votes

$$VTE = \sum_{j=1}^{k} VTP_j + VTI + VTNR + VTN.$$

Now, it is straightforward to estimate the participation in the election (see expression (A.5)), and we use $\widehat{p} = 100(VTE/\widehat{LN})$ where $\widehat{NL}_h = (N_h)(\bar{w}_h) = \frac{N_h}{n_h}\sum_{l=1}^{n_h} w_{h,l}$, and $\widehat{NL} = \sum_{h=1}^{300}\widehat{NL}_h$. In this case, $w_{h,l}$ is the nominal list of the polling station[5] labeled as l from the random sample of federal district h. Here, we mention that the estimator $\widehat{p}$ has better performance than the obvious estimate $\widehat{p}_* = 100(VTE/LN)$.

6. Using the estimates of voting results per district, it is now straightforward to approximate the number of seats by the RM principle, i.e.

$$RM = (\widehat{M}_1, \ldots, \widehat{M}_k, \widehat{NI}).$$

This is simply done by selecting the candidate with more estimated votes as the winner in each district, and then obtaining frequencies by political party. In the case of coalitions, it is imperative to use the coalition agreements to obtain the political party to which the seat would belong (see Sect. A.1).

7. Following the ideas described in Appendix A, it is now possible to obtain

$$\text{CONF} = (NP_1, \ldots, NP_k, \widehat{NI}).$$

Two inputs are needed:

(i) Steps 4, 5 allow us to obtain the vector

$$VNE_d = (VTP_1\,\mathbb{1}(\widehat{q}_1 \geq 3\%), \ldots, VTP_k\,\mathbb{1}(\widehat{q}_k \geq 3\%)),$$

where

$$\widehat{q}_j = 100 \times \left(\frac{VTP_j}{VVE}\right)\%, \text{ for } j = 1, \ldots, k \tag{2}$$

with

$$VVE = \sum_{j=1}^{k} VTP_j + VTI.$$

This is simply the *Valid Vote Cast* (VVE) defined in Sect. A.2. The VVE is important because one of the aims of the Quick Count is to obtain the estimates (2).

(ii) In step 7, the vector RM is obtained.

8. Calculate the error between the estimates and true values using the distances in expressions (1).
9. Steps 3 to 8 can be repeated, e.g. 50,000 times in order approximate the margins of error ϵ_1 and ϵ_2 as in (1). This is done by simply taking the 0.95 sample quantile

[5] A complete register of the Mexican citizens entitled to vote over the polling station.

Table 1 Margins of error at a 95% confidence level for different samples sizes

n- n_h	2012		2015		2018	
	$\hat{\epsilon}_1$	$\hat{\epsilon}_2$	$\hat{\epsilon}_1$	$\hat{\epsilon}_2$	$\hat{\epsilon}_1$	$\hat{\epsilon}_2$
1,500–5	2.9	9	2.8	10	0.7	9
3,000–10	2.3	7	2.4	9	0.5	7
4,500–15	2.0	6	2.3	9	0.4	6
6,000–20	2.0	6	2.2	9	0.4	6
7,500–25	1.7	5	2.0	8	0.3	5
9,000–30	1.7	5	2.0	8	0.3	5

out of the 50,000 measurements obtained via both distances. It is straightforward to change the sample size in step 2, and repeat the process for the new sample size.

To not base our decision only on the information from one election, the same exercise was carried out using also the votes from the 2015 and 218 elections. Table 1 displays the margins of error at a 95% confidence level for various sample sizes, calculated via the average and the maximum number of seats incorrectly classified.

The average number of seats incorrectly classified changes drastically from 2015 to 2018. Instead, the maximum is more stable. Thus, considering the maximum it is clear that there is no big improvement in precision between setting $n_h = 15$ or $n_h = 30$. However, it is essential to obtain estimates between 10 pm and 11:30 pm on June 6. Thus, it would seem that a larger sample size is the best option. Hence, with these considerations it was decided to select a sample of $n_h = 20$ polling stations per strata. This implies an overall sample size of $n = 6,000$ polling stations.

3.2 *Bootstrap Estimation*

Since there is no analytical form to obtain an estimator for the conformation of the Deputy Chamber, we have followed a classic Bootstrap strategy to obtain confidence intervals for the conformation and the percentages (for example, see [3] or [4]). Actually, our Bootstrap strategy is outlined in Sect. 3.2, steps 3 to 7, where the only change is that in step 3 a sample with replacement is chosen from the sample of polling stations. Also, we have set the number of replications to $B = 300$.

4 Incomplete Samples

On election day, polling stations close around 6 pm, then polling officers count the votes and the CAEs in charge of polling stations in the sample start to send information around 7 pm. The voting results of the polling stations are received gradually. The database is updated in real time with the information available. This incomplete sample is sent to the Quick Count committee every 5 min, and is called remittance. The aim of the Quick Count committee is to obtain the final estimate around 10 pm (or as soon as possible). Thus, it is not possible to wait until the full sample arrives because that might never happen, and also at the time of the final estimation, there could be strata with no information. However, to estimate the conformation of the Chamber of Deputies it is necessary to estimate the result of 300 elections to determine the number of deputies by the RM principle. To solve this problem multiple imputation ideas are used; see [9].

4.1 Multiple Imputation

Multiple Imputation (MI) is a three-stage approach. First, missing values are imputed m times using a statistical model based on the available data. Thus, effectively generating m complete datasets. Second, inference is performed on each of the m complete datasets. Third, the m estimates are consolidated/pooled to obtain a single estimate.

MI can be used in cases where the data are missing completely at random (if the events that lead to data being missing are independent both of observable variables and unobservable parameters of interest, and occur entirely at random), missing at random (when the missing values are not random, but these can be fully accounted for by variables where there is complete information), and even when the data are missing not at random (there is a relationship between the propensity of a value to be missing and its values).

5 Election Day: June 6, 2021

5.1 Election Day Strategy

A simple system using the R programming language (see [8]) was coded to obtain estimates with every remittance on election day; this system is described as follows.

1. Every 5 min, the system looks for a new remittance in a previously defined folder.
2. Via MI, generate $m = 15$ complete datasets of $n = 6,345$ polling stations. To perform MI we use the `mice` library of R; see [11] and [10]. The predictive mean matching (pmm) algorithm was used to generate the imputed datasets.
3. Run the bootstrap strategy with $B = 300$ bootstrap re-samples to obtain estimates for the conformation, proportion and participation with each complete dataset. Here, parallel processing was used to reduce the computational time; see the libraries `foreach` ([7]) and `doParallel` ([1]) of R.

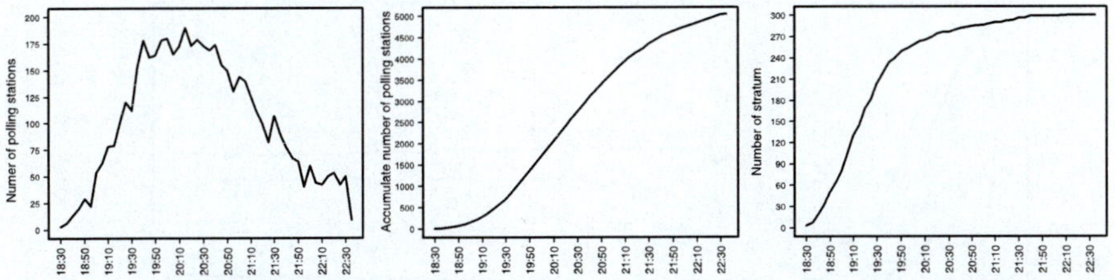

Fig. 1 Arrival of information summaries (time series): number of new polling stations received on each remittance (left), total number of polling stations (center) and number of strata where polling stations have been received

4. Consolidate these $m = 15$ estimates for the conformation of the Deputy Chamber, proportion and participation. These were done following the pooling formulas 3.1.2 to 3.2.10 in [9].
5. Update the time series of estimates.

These five steps take on average 2.5 min, thus it was possible to obtain estimates every 5 min.

5.2 *Arrival of Information*

Figure 1 shows summaries of the number of polling stations received on each remittance.

The first remittance with information was received at 18:30 hrs, and it contained information of only 3 polling stations on 3 different strata. The final remittance was that of the 22:35 hrs with 5,040 polling stations (79.4% of the complete sample of 6,345 polling stations), and information in all the design strata. The average number of polling places per strata was 17, while the minimum and maximum were 3 and 28, respectively.

5.3 *Estimations*

In Fig. 2, the time series for the estimates of the conformation of the Deputy Chamber is presented. The conformation of the Deputy obtained via the district computations of the 11th of June 2021 has been included (dashed lines—true values). An excellent agreement between the interval estimates and the true values is observed in all cases.

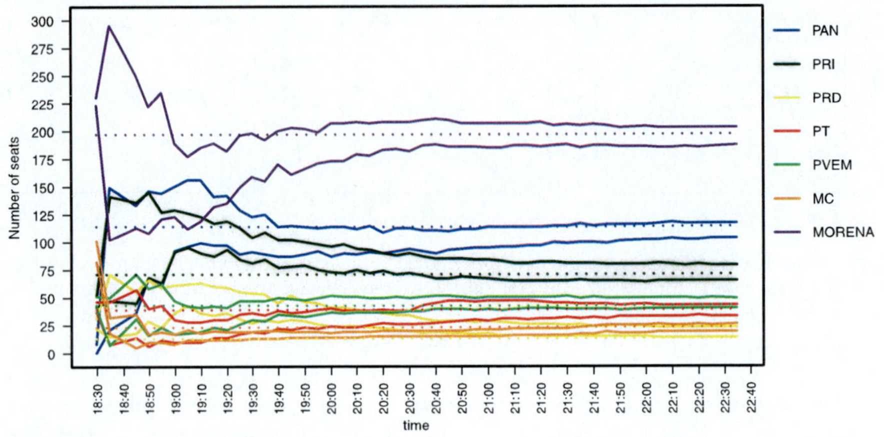

Fig. 2 Time series of estimations for the conformation of the Deputy Chamber, calculated on the evening of the 6th of June 2021 (solid lines), against the conformation obtained via the district computations of the 11th of June 2021 (dashed lines—true values)

6 Discussion

The Quick Count for the Federal Election of Deputies was successful in delivering accurate estimates with a good opportunity. Considering that the INE has adopted the Quick Count as a permanent tool for every federal and local electoral process, it is important to review some of the challenges that we will have to face.

The first challenge concerns the bias in the arrival of the information and that the estimation is always calculated under an incomplete sample setting. The second challenge is to define a sample design that reduces the polling stations that every CAE, that participates in the Quick Count, has to report without sacrificing precision in the estimation. Other problems are, for example, how to address situations when the estimated intervals of two or more candidates intersect with each other; the most efficient strategy to consolidate three or more interval estimates for the same candidate, etc.

A Appendix: Transforming Votes into Seats in the Deputy Chamber

This appendix presents a detailed description of the considerations and procedures for transforming the votes received at the polls into seats in the Deputy Chamber.

A.1 Relative Majority

The territory of Mexico is divided into 300 Federal Electoral districts (article 53 of the [2]). These districts do not intersect with each other and their union is the entire country. The 300 deputies through the RM principle are obtained by holding independent elections in each of these 300 districts. In each district, there can be only one winner, either from a political party or an independent candidate. Then, the vector of seats that each political force obtains through the RM principle is given by

$$RM = (M_1, \ldots, M_k, NI) . \tag{A.1}$$

To obtain the vector RM, it is necessary to consider whether there are coalitions in the districts or not. Under a coalition, to determine the political party to which the seat belongs, the coalition agreements must be taken into account. Also, it is important to consider the INE/CG193/2021 agreement,[6] with regard to effective affiliation and relevance to a parliamentary group. This is described in detail in the next section.

Coalitions

In each of the 300 elections, there may be coalitions between different parties. If a coalition exists, the following considerations must be considered:

- On the ballot, voters can validly choose any combination of parties in the coalition, and the vote will be added to the coalition's candidate. For example, if in the i-th district there is the two-party coalition: $P_{i,1}$ and $P_{i,2}$. Valid votes on the ballot result from any of the following combinations: $\{P_{i,1}\}$, $\{P_{i,2}\}$ and $\{P_{i,1}, P_{i,2}\}$. Any of these possible voting combinations would imply one vote in favor of the coalition's candidate. Similarly, it may happen that in the j-th district, there is a coalition between three parties: $P_{j,1}$, $P_{j,2}$ and $P_{j,3}$. Thus, the valid ways to vote would be selecting any of the combinations $\{P_{j,1}\}$, $\{P_{j,2}\}$, $\{P_{j,3}\}$, $\{P_{j,1}, P_{j,2}\}$, $\{P_{j,1}, P_{j,3}\}$, $\{P_{j,2}, P_{j,3}\}$ and $\{P_{j,1}, P_{j,2}, P_{j,3}\}$. If there are no coalitions in a given district, the valid votes are only the votes for each party directly.
- If there is a coalition in the i-th district, the sum of the votes obtained by all valid ways of voting for the coalition will be used in favor of the coalition candidate. At this point, it is necessary to consider both coalition agreements and effective membership:

  i. The political parties form a coalition and there is only one candidate for the coalition. As already mentioned, votes in favor of any party in the coalition (or valid combination) are considered in favor of this candidate. In addition, in case

---

6 https://repositoriodocumental.ine.mx/xmlui/bitstream/handle/123456789/118249/CG2ex202103-19-ap-3.pdf.

the coalition wins, the parties previously agree on the coalition party for which the seat will be counted (see the original coalition agreements[7]).

ii. Effective membership is a concept that was used for the first time in the 2021 election (see INE/CG193/2021 agreement). The effective affiliation is a validation that the INE makes to the coalition agreements. Here, it is verified that the candidate is a member of the party that indicates the agreement. Otherwise, a correction is made, and the seat is counted to the party in which the coalition candidate is active.

- To determine the weight of each party at the national level, it is important to distribute the coalition votes among the parties that made up the coalition. This distribution is done at the Federal District level. In this case, the votes for valid combinations that include two or more parties are divided between each party in the combination. Individual votes for each party are not shared. For example, in the case of a coalition between the parties $P_{i,1}$ and $P_{i,2}$ in the i-th district. The votes obtained by the combination would have to be distributed $\{P_{i,1}, P_{i,2}\}$, between parties $P_{i,1}$ and $P_{i,2}$. This is achieved through the following procedure:

(a) The total number of votes of the combination is calculated $\{P_{i,1}, P_{i,2}\}$, i.e. $v_1 = \#\text{votes}\{P_{i,1}, P_{i,2}\}$.
(b) The quantity $v_2 = \left\lfloor \frac{v_1}{2} \right\rfloor$ [8] is obtained, as well as the remainder $v_3 = v_1 - 2v_2$ and the votes for each party are calculated considering both direct votes and joint votes, i.e. $VP_{i,1} = \#\text{votes}\{P_{i,1}\} + v_2$ and $VP_{i,2} = \#\text{votes}\{P_{i,2}\} + v_2$.
(c) The remainder is assigned to the party with the most votes, i.e. if $VP_{i,1} \geq VP_{i,2}$, then $VP_{i,1} = VP_{i,1} + v_3$ and in another case $VP_{i,2} = VP_{i,2} + v_3$.

Assuming that in the j-th district there was a coalition between three parties, following an analogous procedure, the votes obtained by the combinations $\{P_{j,1}, P_{j,2}\}$, $\{P_{j,1}, P_{j,3}\}$, $\{P_{j,2}, P_{j,3}\}$ and $\{P_{j,1}, P_{j,2}, P_{j,3}\}$ would have to be distributed between each political party of the combination.

Through the previous procedure, the total votes in favor of each political party in each of the 300 federal districts are obtained, i.e. $VTP_{i,j}$ for $j = 1, \ldots, k$ and $i = 1, \ldots, 300$.

A.2 Voting Types Considered in the Law

The RM principle is based on the number of votes each candidate obtains. Instead, the PR principle depends on several considerations. Before describing the latter, we define three voting types necessary to implement it.

---

[7] https://portal.ine.mx/actores-politicos/convenios-de-coalicion/.

[8] Where for any real number b, we have that $\lfloor b \rfloor$ is the largest integer less than or equal to b. As a consequence $b \geq \lfloor b \rfloor$.

Total Vote Cast

The total vote cast (VTE by its Spanish abbreviation) is the sum of all the votes cast at the polls; see article 15 of the [6]. To obtain the deputies by the PR principle, the first step is to obtain the VTE for each political party at the national level. Obviously, the coalition agreements and effective membership must have been considered before reaching this point. The VTE in the i-th district can be written as follows:

$$VTE_{i,\cdot} = (VTP_{i,1}, \ldots, VTP_{i,k}, VTI_{i,1}, \ldots, VTI_{i,r}, VTNR_i, VTN_i), \quad \text{(A.2)}$$

where $VTP_{i,j}$ is the total number of votes in favor of the j-th political party in the i-th district, $VTI_{i,s}$ is the total number of votes in favor of the s-th independent candidate in the same district, $VTNR_i$ are the votes for unregistered candidates and VTN_i are the null votes.

Finally, the VTE disaggregated by party, independent candidates, votes for unregistered and null candidates is given by

$$VTE_d = (VTP_1, \ldots, VTP_k, VTI_1, \ldots, VTI_r, VTNR, VTN), \quad \text{(A.3)}$$

where $VTP_j = \sum_{h=1}^{300} VTP_{h,j}$; for $j = 1, \ldots, k$ in the same way, to obtain the total votes for independents, unregistered candidates and null votes, these are added over the 300 districts. Finally, the VTE is obtained simply by summing the components of the vector VTE_d, i.e.

$$VTE = \sum_{j=1}^{k} VTP_j + \sum_{s=1}^{r} VTI_s + VTNR + VTN. \quad \text{(A.4)}$$

With this, the citizen participation in the federal election is calculated via

$$p = 100 \left(\frac{VTE}{NL} \right), \quad \text{(A.5)}$$

where NL is simply the total number of people entitled to vote (known as nominal list). In the 2021 election, the size of the nominal list was of $NL = 93,455,692$.

Valid Vote Cast

The *Valid Vote Cast* (VVE by its Spanish abbreviation) is given by the VTE minus invalid votes, that is,

$$VVE = VTE - (VTNR + VTN) = \sum_{j=1}^{k} VTP_j + \sum_{s=1}^{r} VTI_s;$$

see article 15 of the [6] and Sect. II of article 54 of the [2]. The disaggregated VVE is obviously defined as

$$VVE_d = (VTP_1, \ldots, VTP_k, VTI_1, \ldots, VTI_r). \tag{A.6}$$

According to the paragraph added in the Official Gazette of the Federation 02-10-2014 in Sect. I of article 41 of [2], the j-th political party will preserve its registry at the National level, only if the percentage of its total number of votes with respect to VVE is greater than or equal to 3%, i.e. if $q_j \geq 3\%$, then the P_j party retains its registration at the national level, where

$$q_j = 100 \left(\frac{VTP_j}{VVE} \right), \text{ for } j = 1, \ldots, k. \tag{A.7}$$

For this reason, one of the objectives of the Quick Count is to estimate $q_1, \ldots, q_k$. Observe that the sum of these percentages is not 100% since the independent candidates, who are also part of the VVE, are not being considered.

National Vote Cast

The *National Vote Cast* (VNE by its Spanish abbreviation) aims to include in the PR principle only those political parties that have obtained "enough" votes. Hence, the VNE is defined only by the votes of political parties (without independent candidates) and given by the total votes of those parties that comply with having at least 3% of the VVE. The disaggregated VNE is defined as

$$VNE_d = (VTP_1 \, \mathbb{1}(q_1 \geq 3\%), \ldots, VTP_k \, \mathbb{1}(q_k \geq 3\%));$$

see article 15–2 of [6] and Sect. III of article 54 of the [2]. Then, the VNE is obtained as

$$VNE = \sum_{j=1}^{k} VTP_j \, \mathbb{1}(q_j \geq 3\%), \tag{A.8}$$

where $\mathbb{1}_A(x)$ is the indicator function defined as 1 if $x \in A$ and 0 in another case.

A.3 Proportional Representation

The inputs to calculate the 200 deputies by the PR principle are the number of deputies obtained by the RM principle of each political force (A.1), as well as the

VNE_d, (A.8). Based on this information, the 200 deputies are distributed according to the PR principle as described below.

First Distribution

Start by calculating the Natural Quotient (NQ) as $NQ = \frac{VNE}{200}$; see [6], article 16–2. NQ can be interpreted as the number of votes that each seat in the Chamber is worth under the PR principle. To carry out the first distribution, the deputies assigned to each political party will be determined as (see [6], articles 16–3 and 17–1)

$$C_j = \left\lfloor \frac{VTP_j\, \mathbb{1}(q_j \geq 3\%)}{NQ} \right\rfloor, \quad \text{for } j = 1, \ldots, k.$$

Then, it is clear that

$$200 = \frac{VNE}{NQ} = \sum_{j=1}^{k} \frac{VTP_j\, \mathbb{1}(q_j \geq 3\%)}{NQ} \geq \sum_{j=1}^{k} C_j.$$

Therefore, we need to assign $\gamma = 200 - \sum_{j=1}^{k} C_j$ deputies; this is done following the *largest remainder method*:

1. The residuals are calculated, i.e. $\delta_j = VTP_j\, \mathbb{1}(q_j \geq 3\%) - (C_j)(NQ)$, for $j = 1, \ldots, k$.
2. The political parties are sorted from highest to lowest according to their residuals.
3. A seat is assigned to each party according to that order until there are no more deputies to distribute.

It is important to note that the reminder γ is always less or equal to the number of political parties that contribute to the VNE. Therefore, the previous procedure will always lead to the distribution of the 200 deputies, denoting $L_j \in \{0, 1\}$ to the places awarded to the party P_j, for $j = 1, \ldots, k$, by the *largest remainder method*. The total number of deputies via the PR mechanism, which each party has up to this point, is obtained via $R_j = C_j + L_j$, para $j = 1 \ldots, k$, and clearly $\sum_{j=1}^{k} R_j = 200$. Therefore, the total number of deputies that would correspond to the party P_j is given by the sum of deputies obtained via the principles of RM and PR, i.e. $NP_j = M_j + R_j$, for $j = 1, \ldots, k$.

Overrepresentation

In an ideal democracy, the composition of the Chamber of Deputies would faithfully represent each of the sectors of society. In the case of Mexico, the 500 deputies should be distributed proportionally according to the VVE. However, the law allows a limited overrepresentation by establishing two upper bounds:

1. No political party can have more than 300 deputies by both principles (RM and PR).
2. No political party can have a greater number of seats than what results from considering 8% of its participation in the VNE.

The number of seats for the political party P_j satisfies the overrepresentation bounds if

$$NP_j \leq U_j \, \mathbb{1}(q_j \geq 3\%), \quad \text{for } j = 1, \ldots, k, \tag{A.9}$$

where

$$U_j = \min\left\{300, 500\left(\frac{VTP_j}{VNE} + 8\%\right)\right\}.$$

LEGIPE [6], article 17–2, indicates that the party that does not comply with this limit should lose the necessary seats by the PR principle, until the condition is satisfied.

Adjustment

In the event that any political party exceeds the overrepresentation bound, an adjustment is necessary. First, the political party in excess is assigned the maximum number of seats possible via the PR principle (so that it does not exceed the limit (A.9)). Second, this party is removed from the procedure and the remaining seats are distributed among the parties that are not in excess; see article 18–1 of [6]. It is important to observe that these steps imply an iterative process, since (1) there may be several parties that exceed the limit and (2) when re-assigning deputies it is necessary to verify again that the bound (A.9) is satisfied by the remaining parties. The algorithm is as follows:

0. Initialize the indicator variables G_j that will tell us if a political party is (or was) in excess; additionally, parties that do not contribute to the VNE are discarded, i.e. $G_j = 1 - \mathbb{1}(q_j \geq 3\%), \ \text{for } j = 1 \ldots, k.$
1. Calculate the excess for each party, via

$$E_j = (NP_j - U_j)(1 - G_j)\, \mathbb{1}(NP_j > U_j), \quad \text{for } j = 1 \ldots, k.$$

   Clearly, $E_j > 0$, if and only if (1) P_j is one of the parties that contributes to the VNE and has not been in excess and (2) it is in excess in this iteration, i.e. $NP_j > U_j$.
2. Update the G_j variables, i.e.

$$G_j = \begin{cases} 1, & \text{if} \quad E_j > 0, \\ G_j, & \text{otherwise,} \end{cases} \quad \text{for } j = 1 \ldots, k.$$

3. Assign the maximum number of seats possible by the PR principle to the parties that are in excess. This is done simply by removing the surplus, i.e. $R_j = G_j(R_j - E_j), \text{ for } j = 1\ldots, k.$
4. Distribute the remainder of seats, i.e. $\beta = 200 - \sum_{j=1}^{k} R_j$, among the parties that are not (or have not been) in excess. Based on article 18 of [6], the *Effective National Cast* (VNE_f) is calculated by subtracting from the VNE the votes of the parties that are or were in excess, as well as those that are not part of the VNE, i.e.
$$VNE_f = VNE - \sum_{j=1}^{k}(G_j)(VTP_j)\,\mathbb{1}(q_j \geq 3\%).$$
5. A new natural quotient is calculated via $NQ = \frac{VNE_f}{\beta}$, then a procedure analogous to that carried out in the initial distribution is followed. These ϵ seats are distributed, first considering
$$C_j = \left\lfloor \frac{VTP_j\,\mathbb{1}(q_j \geq 3\%)(1 - G_j)}{NQ} \right\rfloor, \quad \text{for } j = 1\ldots, k.$$
Noting that $\gamma = \beta - \sum_{j=1}^{k} C_j$ remains to be assigned, the residuals
$$\delta_j = VTP_j\,\mathbb{1}(q_i \geq 3\%)(1 - G_j) - (C_j)(NQ), \quad \text{for } j = 1, \ldots, k,$$
are calculated. Via the *largest remainder method*, $L_j \in \{0, 1\}$ is obtained and the number of seats by the PR principle is updated calculating
$$R_j = R_j + C_j + L_j, \quad \text{for } j = 1, \ldots, k.$$
Note for some j, $C_j = 0$ (thus $L_j = 0$), hence $R_j = R_j$ is a consistent step.
6. Update the number of seats, i.e. $NP_j = M_j + R_j$.
7. The excess is calculated according to step 2; here, there are two options: (i) if there is a party in excess, proceed to step 3, and continue with the remaining steps; (ii) in case there are no more parties in excess, the algorithm ends (see article 17–2 of the [6]).

References

1. Corporation, M., Weston, S.: doParallel: Foreach Parallel Adaptor for the 'Parallel' Package. R package version 1.0.16 (2020) https://CRAN.R-project.org/package=doParallel
2. CPEUM.: Constitución Política de los Estados Unidos Mexicanos (2021). http://www.diputados.gob.mx/LeyesBiblio/pdf_mov/Constitucion_Politica.pdf
3. Efron, B.: Bootstrap methods: another look at the jackknife. Ann. Stat. **7**(1), 1–26 (1979)

4. Efron, B., Tibshirani, R.: An Introduction to the Bootstrap. Chapman & Hall/CRC. Monographs on Statistics & Applied Probability. Taylor & Francis (1993)
5. Gilas, K.M., Medina Torres, L.E.: Asignación de diputaciones por el principio de representación proporcional. TEPJF, Ciudad de México, México (2012)
6. LEGIPE.: Ley general de instituciones y procedimientos electorales (2017). https://www.ine.mx/wp-content/uploads/2020/07/Despen-LEGIPE-NormaINE.pdf
7. Microsoft and Weston, S.: Foreach: Provides Foreach Looping Construct. R package version 1.5.1 (2020). https://CRAN.R-project.org/package=foreach
8. R Core Team.: R: A Language and Environment for Statistical Computing, R Foundation for Statistical Computing, Vienna, Austria (2021). https://www.R-project.org/
9. Rubin, D.B.: Multiple Imputation for Nonresponse in Surveys, Wiley Series in Probability and Statistics. John Wiley & Sons Inc (1987)
10. van Buuren, S.: Flexible Imputation of Missing Data, 2nd edn. Chapman & Hall/CRC (2018)
11. van Buuren, S., Groothuis-Oudshoorn, K.: MICE: multivariate imputation by chained equations. R J. Stat. Softw. **3**(45), 1–6 (2011)

Bayesian Analysis of Homicide Rates in Mexico from 2000 to 2012

María de los Dolores Sánchez-Castañeda, Luis E. Nieto-Barajas, and María del Pilar Fuerte Celis

1 Introduction

Homicides in Mexico showed a downward trend since 1992. However, it drastically reverted to an upward trend since the declaration of the "war against drugs" in late 2006 reaching unprecedented levels as a result of territorial confrontations of drug cartels against army forces and against other cartels [2, 4]. Between 2006 and 2010, there were 34,550 homicides related to organized crime, compared to 8,901 cases registered between 2000 and 2006 [27]. In 2012, the then new government decided to change the public security strategy and created a National Program for Social Prevention of Violence and Delinquency. This decision acknowledges social prevention as the main axis of national security policy.

Despite great efforts to analyze spatial patterns of dispersion of homicide in Mexico [11, 14, 24], the study of specific mechanisms still is at an early stage [8, 16]. A statistical and geographic approach is needed to investigate common features of geographic closeness of homicides and demographic variables of victims during 2000–2012, corresponding to the period of the war on drugs.

It is important to create better indices to develop efficient prevention and fight policies against violence and crime. This can be achieved through analysis of data related to violence, the places where they happen, their demographic features, and their evolution in time. By identifying regions with the highest homicide rates by sex and age, and their evolution in time, policy makers would have valuable information about vulnerable groups and use it to design crime prevention policies, and budget

M. de los D. Sánchez-Castañeda (✉)
Banco de México, Subgerencia de Análisis Estadístico, CDMX, Mexico City 06000, Mexico
e-mail: mariasc@banxico.org.mx

L. E. Nieto-Barajas
ITAM, Department of Statistics, CDMX, Mexico City 01080, Mexico

M. del P. Fuerte Celis
Research Center for Geospatial Information Sciences, Mexico City 14240, Mexico

I. Antoniano-Villalobos et al. (eds.), *Interdisciplinary Statistics in Mexico*,
Springer Proceedings in Mathematics & Statistics 397,
https://doi.org/10.1007/978-3-031-12778-6_13

allocation targeted at the regional level that may produce more effective results than general policies at the national level. A predictive model of homicide rates can provide such information.

Hence, the aim of this paper is to fit a spatial model at the state level of annual homicide rates in Mexico from 2000 to 2012 by age and gender using Bayesian estimation and spatial distributions. By including spatial and time dependence, it is possible to identify whether homicide rates have a contagious effect on neighboring states over time. Moreover, by standardizing the rates by age and gender, the model can reveal vulnerable subpopulations. The model estimates parameters of main effects of four covariates: year, state, age, and sex; it also estimates two-way interactions: time-state, state-sex, sex-age, and time-sex. Since covariates are categorical, we use a factorial model. Hence, we refer to effects as in factorial models with fixed or random effects, where one investigates the effect of a factor on the response variable, not as causal effects or causal inference.

We use data of homicide counts from 2000 to 2012 to fit a Bayesian Poisson regression model with conditionally autoregressive (CAR) process prior that is appropriate to model spatial effects and spatial interactions, for annual temporal effects and time interactions we use a dynamic linear prior, and independent priors for sex and 5-year age-group effects.

Our results are consistent in detecting specific moments and places like in Tamaulipas where a massive burial was found in 2012, coinciding with a high effect of interaction state-year from 2010 to 2012. Also, point estimates of interaction of sex-year detect an overall increasing risk of female homicides from 2001 to 2007; the effect was reverted due to the increase of male homicide after the war on drugs strategy was declared.

The paper has the following structure. Section 2 presents a brief literature review of homicide analysis and modeling. Section 3 presents an exploratory data description. Section 4 provides the methodological framework of the model. Section 5 presents the results. Finally, Sect. 6 presents the main findings and discussion of this work.

2 Literature Review

Analysis of homicide can be approached from different areas. In Sociology, it is considered that homicide is related to violence as a visible manifestation of physical aggression where an individual exerts by force the will to power over another [26]. Violence is considered as a historical and social process whose origins are a series of individual and social determinants that constitute an action causing more than just death. Each homicide is a consequence of an individual decision of the perpetrator; it obeys accidental motivations, each with its own explanation, but it is not entirely random. If we analyze homicides in Mexico during a given period of time, the distribution is not completely random. There are some regions with concentration patterns. These patterns show fundamental elements of the social order.

Research on homicides in Mexico is approached from two perspectives, the journalistic and the academic perspective. Each one addresses different aspects of the problem, from relations of power, drug trafficking organizations to the consequences of them. Most studies consider territory as a determining factor. Snyder and Durán-Martínez [28], Durán-Martínez [9], Ríos Contreras [27], and Dell [8] analyze the political factors that cause violence. Guerrero-Gutierrez [14] investigates the effect of law enforcement on the conflict. García-Ponce et al. [13] and Castillo et al. [3] analyze the availability of weapons as possible causes of an increase in homicides.

Quantitative studies of homicide explore the impact of violence on economic variables like investment, economic growth and development at state level, [25], on human capital [22], and on employment rates [20].

Other works study the spatial dispersion of homicide. Osorio [24] proposes a game-theoretic intensity model of competition for territorial dominance between drug cartels at a municipal level during 2000–2012. Espinal-Enriquez and Larralde [10] analyze the diffusion of homicides in Mexican municipalities using a network model to identify those with the highest number of interactions as centers where criminal organizations are based. Curiel et al. [6] use a statistical approach to propose a measure of the distribution and concentration of crime in Mexico City using a mixture model. Also, [12] use spatial association statistics like Moran's I and Getis-Ord tests on homicide rates, population density, migration intensity rate, and social gap index, among other variables to find clusters of municipalities that share features related to violence within them, but different among each other during 2000 to 2012. They use random forest classification for variable selection, variable importance, and to validate clusters

Among studies that use spatio-temporal models to estimate the effect of demographic variables on homicides in Mexico we found two. Leenen and Cervantes-Trejo [19] analyze spatio-temporal trends of homicide and suicide rates from 1998 to 2012; they fit a spatial logistic regression model by sex and 10-year age-groups by municipalities. However, they did not consider spatial dependence. Only the work of [17] was found to use Bayesian estimation. They use a two-level spatial lagged linear model to estimate homicide spillover effects of sex and age-group in Mexico between 2005 and 2010. They included other proxies to variables related to violence like distance to the U.S. border, drug trafficking activities, and military interventions. This model did not consider spatial distributions.

In [12], the authors consider mainly social variables, but to get a better understanding of the phenomenon, we also need to find in more detail the demographic characteristics of the population vulnerable to homicide from a spatio-temporal perspective. Hence, the main contribution of this work is to estimate spatial main effects, and spatial interactions with sex, age, and time using Bayesian estimation of spatial distributions specified as conditional autoregressive priors.

3 Data

The data used are official records of homicides from the National Statistics Bureau of Mexico (INEGI), from years 2000 to 2012. Records consist of individual cases of homicides by day-month-year of event, state and municipality of occurrence, sex and age of the victim, type of homicide, and place of occurrence (highway, street, office, and home), among other variables. The data are aggregated by variables given by the cross product of the following categories: sex (male or female), quinquennial age-group (0–4, 5–9, ..., 95–99, 100+), year (2000, ..., 2012), and state (ID-1, ..., ID-32).

Homicide rates are expressed with respect to the population size per 100,000 inhabitants. Hence, the population was aggregated at the same level as homicides, i.e. sex, quinquennial age-group, year, and state. Population data is also available from INEGI for the censuses of the years 2000 and 2010, and for the intercensal survey of 2005. Population for intercensal years 2001 to 2004, 2006 to 2009, and 2011 and 2012 are given only by forecasts provided by the Mexican Population Council (CONAPO). For 2011 and 2012, population projections are given by 15-year groups of age, so we apply the population's average structure of the 2000 census to estimate the population by 5-year groups of age. Although the type of homicide and place of occurrence (street, highway, home, etc.) are available, we chose not to include them in the model due to the high dimensionality that it would imply.

Total cases of homicides considered for the 13 year period are 184,918. Records with missing sex or age-group were not considered (6,042 cases). Figure 1 illustrates annual total number of homicides (solid line), and male, and female homicides (dotted and dashed lines, respectively).

Figure 1 clearly shows a decreasing trend in homicide from 2000 to 2007 and a rapid increase from 2008. There was an increase of 60.77% from 2007 to 2008. In 2011, homicide reached the highest level in the 13-year period, with 25,321 cases. Male homicides represent 88.85% of the total, while female homicides represent 11.15%. The distribution of the percentage of homicide by state presented in Fig. 2 shows that nine states concentrate 65.4% of homicides. The states of Mexico and Chihuahua have the highest percentages with 13.06% and 13%, respectively. While Guerrero, Mexico City, and Sinaloa concentrate 20%, followed by Jalisco, Michoacan, and Baja California with 15% of homicide, Yucatan, Aguascalientes, and Baja California Sur have the lowest percentages.

Figure 3 illustrates the pyramid of homicide by 5-year groups of age and sex. The structure of the number of homicides by age-group are similar for both sexes. The groups with the highest percentage of homicides are between 20 and 39 years. Male victims of these groups concentrate 57.94% of cases, while female victims concentrate 56.85% of cases. The groups of 15–44 years concentrate 72.4% of homicides.

Even though the number of male homicides is considerably greater than for females, it is important to note that the female age-group with the highest percentage of homicide is the 20 to 24 years old, whereas for males it is the 25 to 29 years old. This could be indicative of the nature of homicide in this subpopulations; they

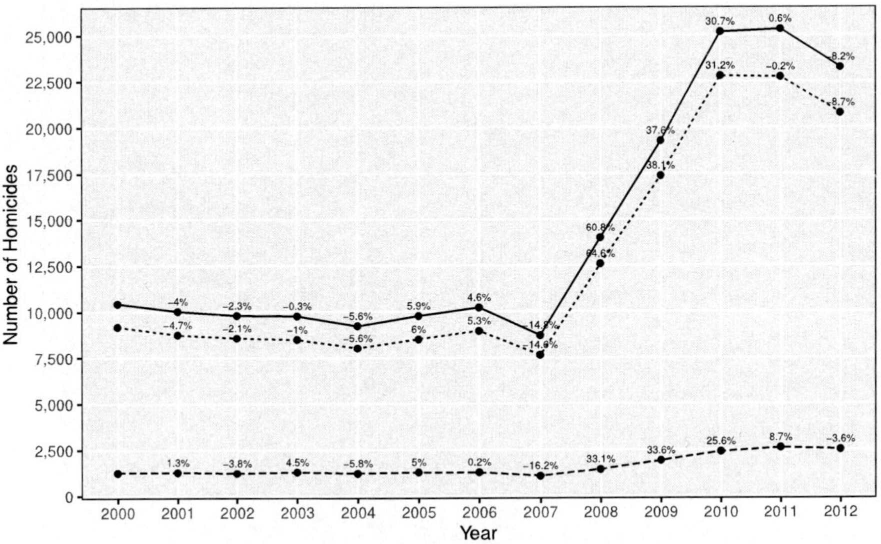

Fig. 1 Annual number of homicides in Mexico 2000–2012. Total (solid line), males (dotted line), and females (dashed line). Annual increase is shown above each line

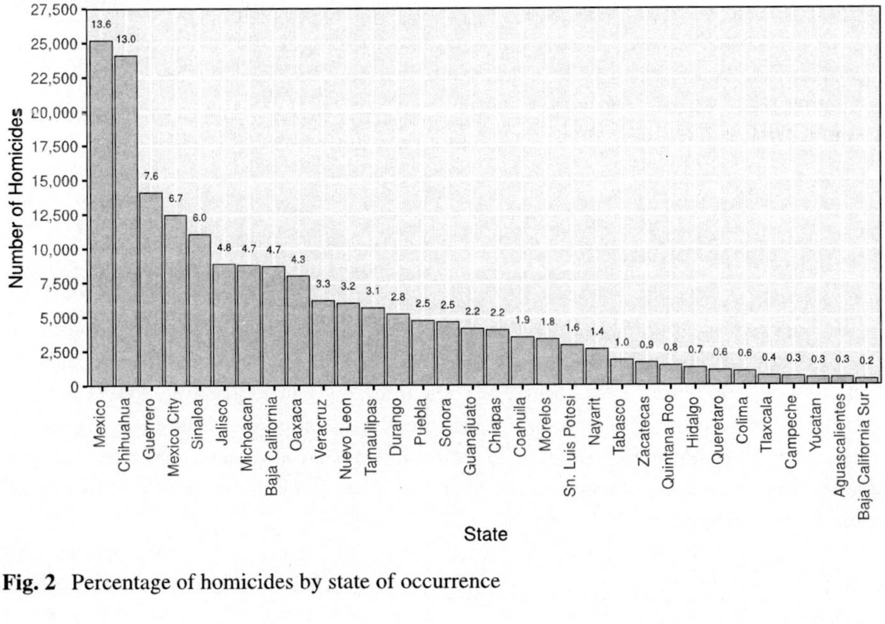

Fig. 2 Percentage of homicides by state of occurrence

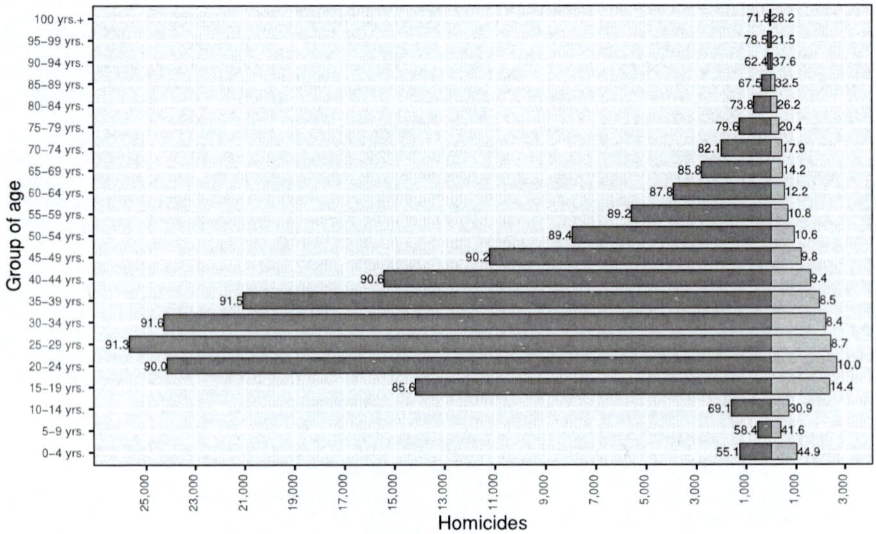

Fig. 3 Pyramid of Homicide by 5-year groups of age and sex. Male cases (left panel) and female cases (right panel)

may be associated with gender violence against women of younger ages. Municipalities like Ciudad Juárez in Chihuahua, and Ecatepec in State of Mexico reported a large number of female homicides leading to a new legal classification of homicide: femicide, that is the homicide of women because they are women or girls.

4 Model Specification

In the analysis of count data, the most common choice of model is Poisson. This model is appropriate when the count of cases is low and the population is large in each observational unit [18]. Since we are interested in estimating the effects of sex, age-group, year, and state as factors on the number of homicides, we use a Poisson regression model with a logarithmic link function to associate covariates to observations by means of a linear function. We also estimate some interactions that could be of interest.

Let $Y_{i,j,t,k}$ be the number of homicides per sex i, age-group j that occurred in year t, and state k, for $i = 1, 2$, $j = 1, \ldots, 21$, $t = 1, \ldots, 13$, and $k = 1, \ldots, 32$. Then we assume

$$Y_{i,j,t,k} \mid \mu_{i,j,t,k} \overset{\text{ind}}{\sim} \text{Po}(\mu_{i,j,t,k}), \tag{1}$$

where

$$\mu_{i,j,t,k} = \lambda_{i,j,t,k} \frac{N_{i,j,t,k}}{100{,}000},$$

with $N_{i,j,t,k}$ the population size, and $\lambda_{i,j,t,k}$ becomes the homicide rate per one hundred thousand inhabitants. Since all explanatory variables (sex, age, time, and state) are categorical, i.e. factors, we will use the notation of factorial models [23] and define homicide rates using a logarithmic link function:

$$\log(\lambda_{i,j,t,k}) = \alpha + \gamma_i + \delta_j + \eta_t + \phi_k + \zeta_{i,j} + \nu_{i,t} + \theta_{i,k} + \beta_{t,k}, \tag{2}$$

where α is the overall mean level; the main factors are sex γ_i, age-group δ_j, year η_t, and state ϕ_k. Interactions of interest are sex-age $\zeta_{i,j}$, sex-year $\nu_{i,t}$, sex-state $\theta_{i,k}$, and year-state $\beta_{t,k}$. As in a factorial model, estimability constraints are needed for model parameters. These are

$$\sum_{i=1}^{2}\gamma_i = \sum_{j=1}^{21}\delta_j = \sum_{t=1}^{13}\eta_t = \sum_{k=1}^{32}\phi_k = 0$$
$$\sum_{i=1}^{2}\sum_{j=1}^{21}\zeta_{i,j} = \sum_{i=1}^{2}\sum_{t=1}^{13}\nu_{i,t} = \sum_{i=1}^{2}\sum_{k=1}^{32}\theta_{i,k} = \sum_{t=1}^{13}\sum_{k=1}^{32}\beta_{t,k} = 0. \tag{3}$$

Bayesian estimation accounts for uncertainty about the parameters. Moreover, instead of using spatial autocorrelation, Bayesian models allow specification of appropriate spatial prior distributions that account for geographical dependence through random effects and hierarchical model specification. There is no need for a closed form of the posterior distribution, given that complex models with high dimensionality and complex prior specification can be estimated in an efficient way with Bayesian computational methods using MCMC, which nowadays are widely available and efficient.

The Bayesian model is completely specified by assigning prior distributions for each parameter. Priors assigned to the overall level, main effects of sex and age-group, and sex-age interaction are independent normal priors with small precision, i.e. $\alpha \sim \mathrm{N}(0, 0.01)$, $\gamma_i \sim \mathrm{N}(0, 0.01)$, $\delta_j \sim \mathrm{N}(0, 0.01)$, and $\zeta_{i,j} \sim \mathrm{N}(0, 0.01)$. For the temporal (year) effect and sex-year interaction, we use dynamic linear priors of the form $\eta_t \mid \eta_{t-1} \sim \mathrm{N}(\eta_{t-1}, \tau_\eta)$ with initial parameter $\eta_1 \sim \mathrm{N}(0, 0.01)$, and $\nu_{i,t} \mid \nu_{i,t-1} \sim \mathrm{N}(\nu_{i,t-1}, \tau_\nu)$ with initial parameter $\nu_{i,1} \sim \mathrm{N}(0, 0.01)$.

For spatial (state) effect and interactions sex-state and year-state, we use conditionally autoregressive (CAR) priors [5]. This is for $\phi = (\phi_1, \ldots, \phi_{32})$, $\theta_i = (\theta_{i,1}, \ldots, \theta_{i,32})$, and $\beta_t = (\beta_{t,1}, \ldots, \beta_{t,32})$; we assign $\phi \sim \mathrm{CAR}(\rho_\phi, \tau_\phi)$, $\theta_i \sim \mathrm{CAR}(\rho_{\theta_i}, \tau_{\theta_i})$, and $\beta_t \sim \mathrm{CAR}(\rho_{\beta_t}, \tau_{\beta_t})$, where $\mathrm{CAR}(\rho, \tau)$ is in fact a multivariate normal distribution with association parameters $\rho \in (0, 1)$ and precision $\tau > 0$, defined through conditional distributions of the form $\phi_k \mid (\phi_r, r \neq k) \sim \mathrm{N}\left(\frac{\rho}{w_{k+}} \sum_{r\neq k} w_{k,r}\phi_r, \tau w_{k+}\right)$, where $w_{k,r} = I(k \sim r)$ indicates whether state k and r are neighbors, and w_{k+} denotes the total number of neighbors of state k. We assign hyper-prior distributions to the association and precision parameters of the form $\rho_\phi \sim \mathrm{Un}(0, 1)$, $\rho_{\theta_i} \sim \mathrm{Un}(0, 1)$, and $\rho_{\beta_t} \sim \mathrm{Un}(0, 1)$, and hierarchical priors for precisions $\tau_\phi \mid \pi \sim \text{Ga}(1, \pi)$, $\tau_{\theta_i} \mid \pi \sim \text{Ga}(1, \pi)$, and $\tau_{\beta_t} \mid \pi \sim \text{Ga}(1, \pi)$ with $\pi \sim \text{Ga}(0.1, 0.1)$.

Estimates are not sensitive to the choice of precision hyperparameters. Estimation was tested for precision values of 0.01 and 0.001 showing robust results in both cases, but a slightly better convergence behavior for 0.01.

Posterior inference of the model was obtained by implementing MCMC in OpenBUGS software [21] via R using the library R2OpenBUGS [29], running 10,000 iterations with a burn-in of 1,000. There are 617 parameters estimated and 36 hyperparameters. The estimability constraints (3) were appropriately incorporated into the BUGS code.

We assess the model in terms of convergence of estimates of the Markov chains, and of the goodness of fit. We assess convergence by looking at the trace plots and ergodic means of the Markov Chains. In terms of goodness of fit, we use three different measures. The deviance information criterion (DIC), the mean squared error (MSE), and the L-measure introduced by [15] that measures a trade-off between variance and bias, and it is estimated for the complete data set. The lower the value of these measures, the better the fit of the model. Measures of goodness of fit were computed for a baseline model that considers only the main effects without interactions, a model with only two interactions, and for the complete model with interactions as specified in Eq. (2). For all three types of models, we used CAR and ICAR prior specification.

5 Analysis and Results

Convergence of the model is achieved for all the parameters when using 5,000 iterations, and it is preserved when varying the number of iterations to 10,000. The three measures of fit are lower for CAR prior specifications compared to ICAR priors. Hence, the following results are presented only for CAR prior specification. The DIC for the baseline model is 99,020 compared to 62,460 of the complete model, with interactions specified in Eq. (2). The MSE is 1,095.50 for the baseline model compared to 641.57 for the complete model, and the L-measure is 289.33 for the baseline model, and 63.32 for the complete model. From these values, we confirm that a complete model with CAR priors has a better fit than the baseline model or incomplete models with either CAR or ICAR priors.

Posterior estimates of parameters are summarized as follows. Point estimates are given by the posterior mean, and 95% credible intervals (CI) are given by the corresponding posterior quantiles 2.5 and 97.5%. For the overall level, we have $\widehat{\alpha} = 1.78$ with a 95% CI (1.75, 1.81). This means that the average homicide rate in the country is around 6 homicides per 100,000 inhabitants. Sex effect estimate is $\widehat{\gamma}_1 = 0.78$ for males and $\widehat{\gamma}_2 = -0.78$ for females with 95% CI of (0.75, 0.80) and (−0.80, −0.75), respectively. This implies that just for being a man, homicide rate increases by 118%, whereas for women the rate reduces by 55% with respect to the overall mean.

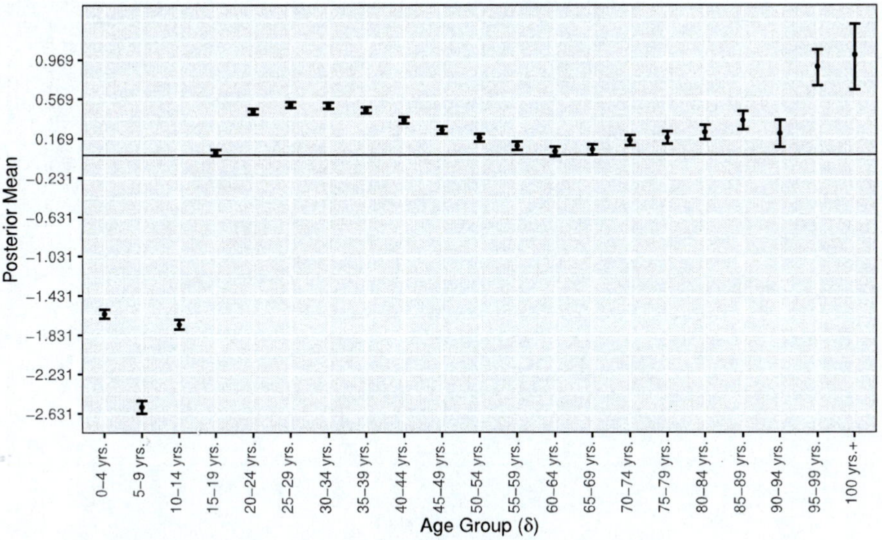

Fig. 4 95% Posterior credible interval of age-group effect δ_j

The age-group effect estimates $\widehat{\delta}$, and their 95% CI are presented in Fig. 4. Risk levels of homicides are related to age, especially for men between 15 and 25 years old because they are exposed to or are more likely to be involved in dangerous activities. At older ages when men have families who depend economically on them, or when their physical condition does not allow them to continue, they decrease or stop doing these activities. However, the eldest groups of 95–99 and 100+ years show the highest estimates. This is because we standardize the effects with respect to the exposed population. The population size of the eldest groups of age is small, and has a small number of homicides; this will result in a greater change in risk of homicide and in greater uncertainty of interval estimates. Hence, these groups of age have the highest and most uncertain estimates. This happens not only for groups of age, but also for factors where the population size is small, for example, state effects of Aguascalientes, Baja California Sur, and Colima or interactions of state-sex for males or females of Aguascalientes, Baja California Sur, or Yucatan. Other age-groups with large effects are 20–24, 25–29, 30–34, and 35–39 age-groups. The rate of homicide increases around 60% for the 20 year-range. For the first three age-groups 0–4, 5–9, and 10–14 years, the effect in homicide rates is negative, that is, the rate decreases 80% for the first and third groups, and 92% for the second group.

Posterior point estimates of calendar year effects $\widehat{\eta}$, and 95% CI are presented in Fig. 5. Time effects are negative for the years 2000–2007. The year with the smallest rate is 2007 with a reduction of 26%, while 2008 shows no differentiated effect with respect to the overall rate. However, higher effects are expected in interactions of spatial effect with other factors. The rest of the years show positive effects, 23% increment in 2009, 42% increment in 2010, 62% increment in 2011, and 55%

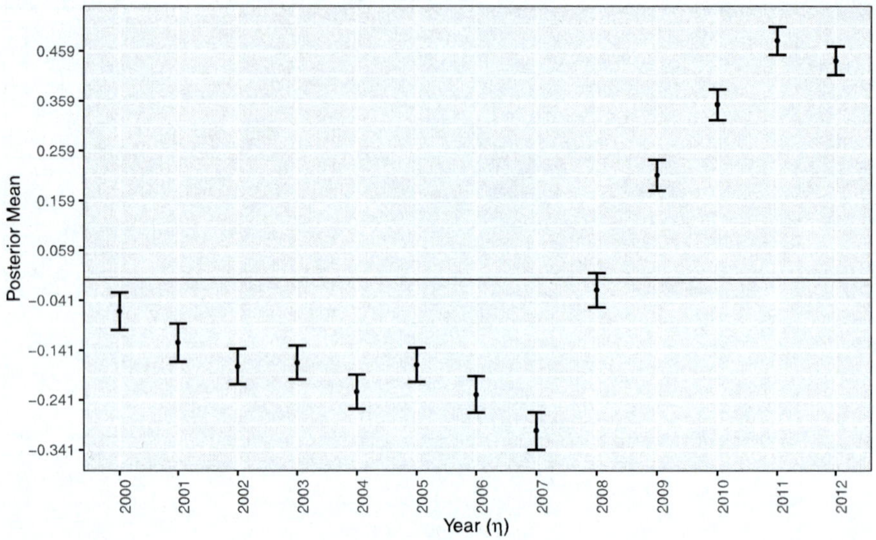

Fig. 5 95% Posterior credible interval of year effects η_t

increment in 2012. This confirms that 2011 has the highest rate reaching 9.2 homicides per 100,000 inhabitants.

Figure 6 shows the heat map of the posterior means for spatial effects $\widehat{\phi}$. The largest effects correspond to Chihuahua and Guerrero with an increase of around 215% with respect to the overall homicide rate. Sinaloa, Baja California, Oaxaca, and Durango come in second place with an increase of around 100%, whereas Nayarit, Michoacán, and Mexico follow in a close third place with an increase of around 60%. These states lie on the Pacific Coast; with the exception of Oaxaca, they are either the main producers of cannabis or poppy, or are in a trafficking route to the United States. The causes of the high effect in Oaxaca might be confrontations over territory limits between communities or protests, and confrontations between teacher unions and police forces that are common in the state. In the opposite direction, the state with the highest reduction is Yucatan with a reduction of 75%, followed closely by Aguascalientes, and Hidalgo with a reduction of around 55%.

Posterior point estimates and 95% CI of interaction effect of sex-age, $\widehat{\zeta}$ are presented in Fig. 7. The left panel corresponds to males, where the age-groups with higher rates go from 15 to 69 years, with a maximum increase of around 50% in the age-group 30-34, whereas for females the highest rates are for the very young groups of age (from 0 to 14 years old) with an increase of 100%, and for the eldest (from 75 to 94 years old) with a maximum increase of 50%.

Posterior point and interval estimates of sex-year interaction, $\widehat{\nu}$ are presented in Fig. 8. In general, we showed that males have a higher rate than females. However, the interaction of sex with calendar year shows that females (right panel) present

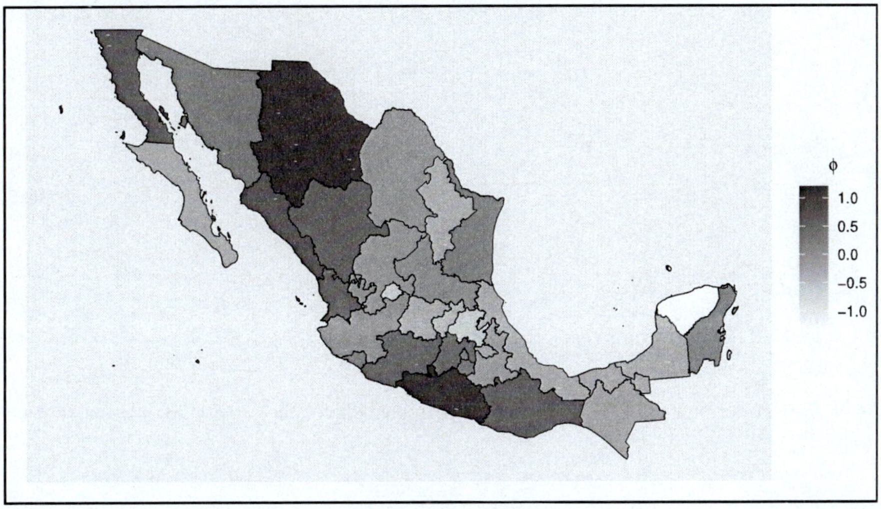

Fig. 6 Map of posterior means of state effects (ϕ)

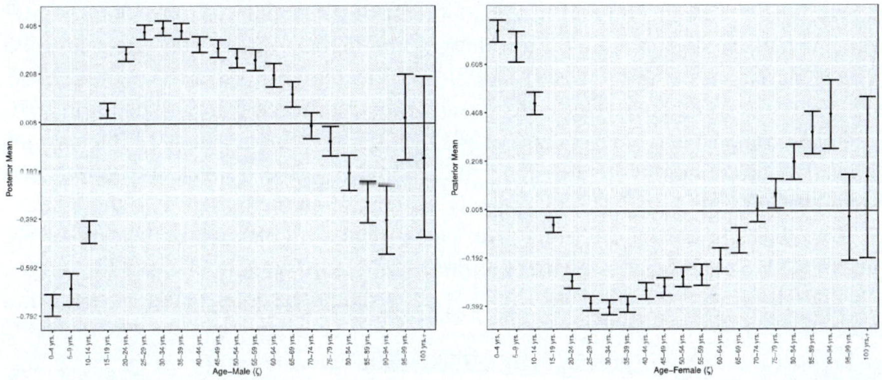

Fig. 7 95% Posterior credible intervals for sex-age interaction effects, ζ. Males (left) and females (right)

an increment of around 5% in homicide rates for the years 2001 to 2007. While for males, there is an increment of around 6% (left panel) for 2008 until 2012.

Point estimates of sex-state interaction $\widehat{\theta}$, and 95% posterior credible intervals are illustrated in Fig. 9. On top of the main effects, these interactions show that for males, states with the highest estimates are Sinaloa (45% increment), Durango (25% increment), Michoacán (20% increment), and Guerrero (20% increment). For females, states with the highest interaction estimates are Tlaxcala (32% increment), Baja California Sur (27% increment), Quintana Roo (27% increment), and Aguascalientes (23% increment). Although states like Chihuahua and Guerrero were found to show the largest effects in homicide rates, no particular sex bias has been detected.

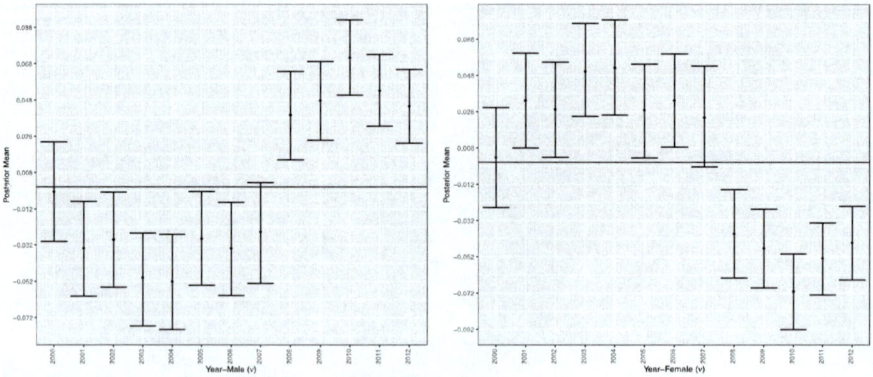

Fig. 8 95% Posterior credible intervals for sex-year interaction effects, ν. Males (left) and females (right)

Finally, estimates of spatio-temporal (year-state) interactions, $\widehat{\beta}$ are shown in Fig. 10 as a panel of annual heat maps. Broadly speaking, from 2000 to 2006 higher interactions were concentrated in the southeast of the country. From 2006 to 2007 Chiapas had an abrupt change, from having the highest interaction effect in 2006 (100% increment) to the lowest in 2007 (50% decrement). From 2008 until 2011, high year-state interactions shifted to the northwest, particularly in Chihuahua in 2010 (250% increment) and Nuevo León in 2011 (300% increment). In 2012 interactions showed that high homicide rates concentrated in the northern states of Coahuila and Nuevo León (both with 200% increment), which are on the border with the US.

Note that since we are dealing with categorical variables whose categories are interrelated to each other, we cannot remove non-significant effects from the model. Effects whose interval estimates include zero should be interpreted as being not different from the overall average effect $\widehat{\alpha}$.

Figures illustrating specific main effects with interactions for all states or heat maps of homicide rates for all sex-age-year-state are not included due to space constraints.

6 Concluding Remarks

To prevent and combat violence in a society, we need to understand the multiple dimensions of violence, its manifestations, causes, impact, and evolution. The multiple dimensions can be classified into four fields, which are economic development, demography, local geography and weapons, trafficking, and organized crime. Homicide is a manifestation of violence and an expression of other social phenomena; it is often used as a proxy variable of violence.

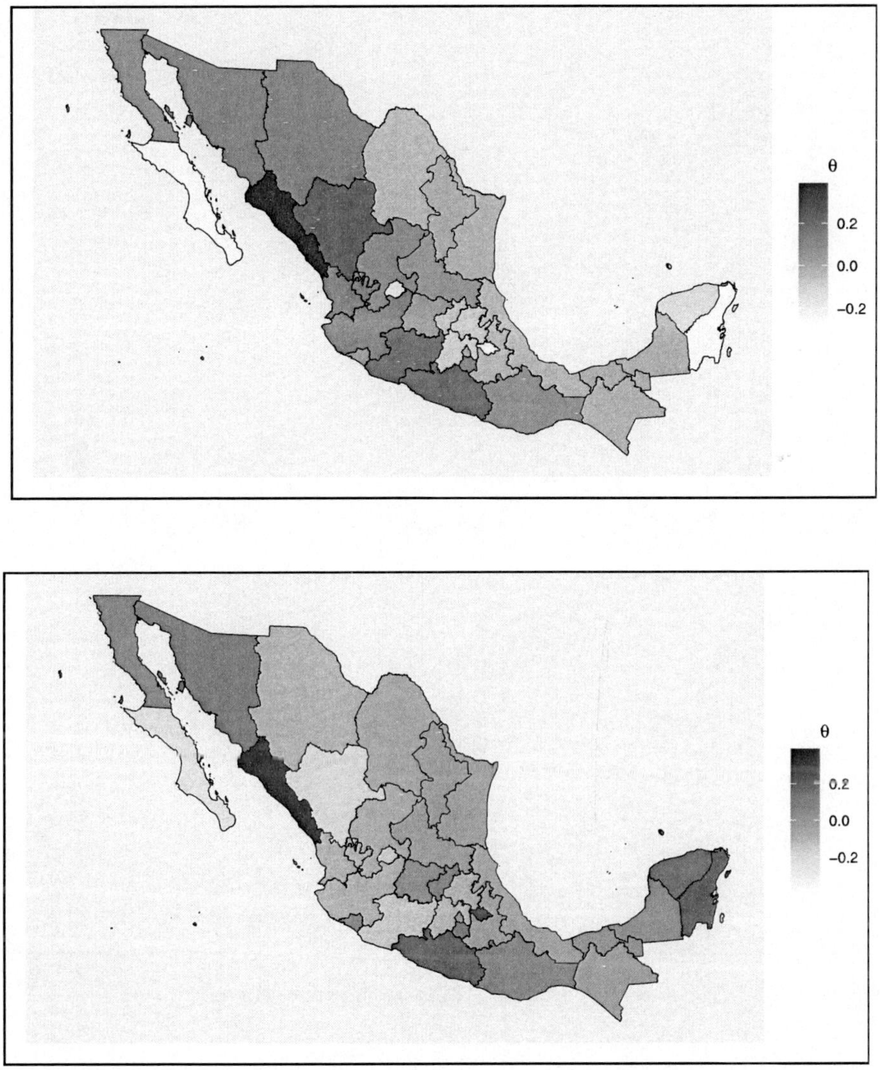

Fig. 9 Posterior means of sex-state interaction effects, θ. Male (above) and female (below)

This paper does not address the economic or social causes of homicide crime in Mexico, but it is based on the assumption that the increase of homicide in Mexico since 2008 is closely related to increased activity of crime organizations and their geographical spread across Mexican states, along with the military strategy that started to combat them in 2007. Far from observing a decrease in violence, the conflict between crime organizations and the army, and between themselves escalated violence, homicide, and other types of crime.

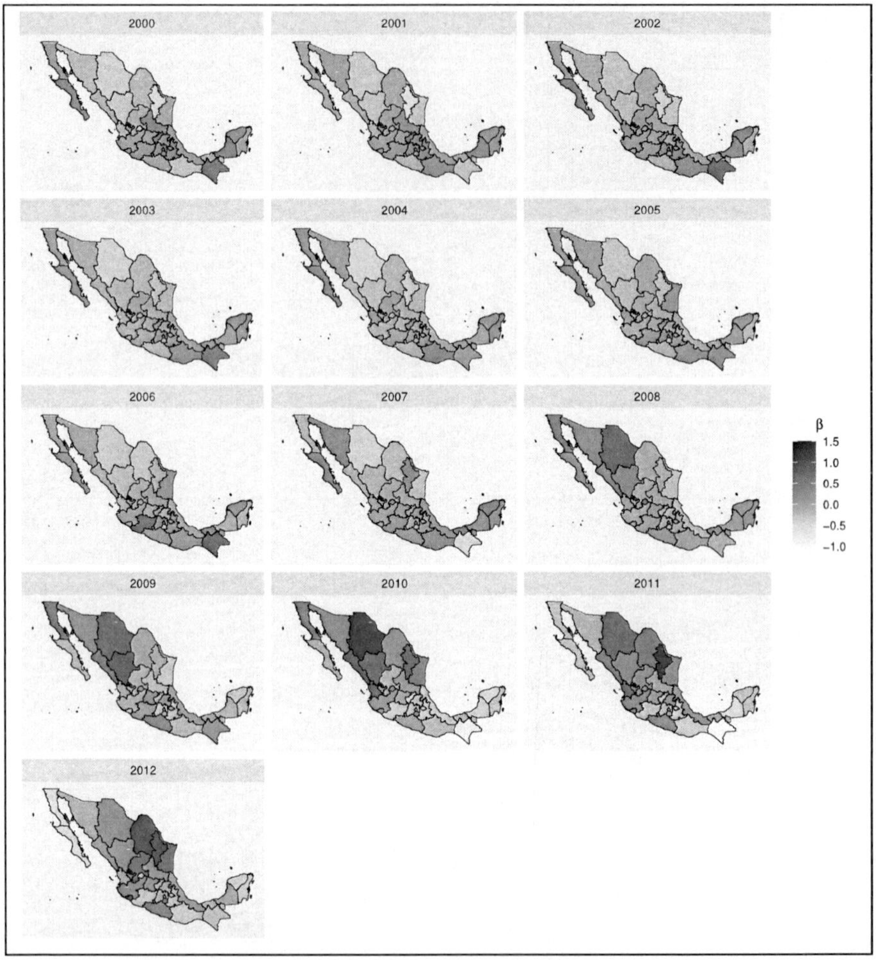

Fig. 10 Posterior means of spatio-temporal (year-state) interaction effects, β

A historical and qualitative approach is important for understanding violence and homicide, also a quantitative approach can provide other tools and guidelines for the design and continuous evaluation of public policies addressed to specific demographic groups.

This article contributes to this topic from a quantitative statistical approach by estimating impacts of demographic and spatial factors on homicide rates, including relevant interactions to avoid confounding factors, and the evolution of these effects in time. Bayesian estimation provides appropriate model specification of complex models with dynamic and spatial dependence in the form of random effects. CAR models are widely used as spatial prior specification that accounts for effects of neighboring units for aereal data, like homicides aggregated by state. Computational

MCMC methods are efficient and readily available for estimating these type of models.

Main effect estimates show that male homicide between ages 20 and 44 years old is significantly positive, and also that effects over time increased drastically from 2008 to 2011, and spatial (state) effects showed that states in the south and center presented higher effects.

Large state effect estimates are consistent with the confirmed presence of drug trafficking, and crime organizations. Some of them occur in Sinaloa and Durango with the presence of the Sinaloa cartel [1], and consequently had military interventions. Also, Chihuahua has the presence of the Juárez cartel; it is on the border with the US, and has both legal and illegal crossing points where human and drug traffic takes place. Since 1993, Ciudad Juárez in Chihuahua has suffered a wave of femicides with extreme violence [7], but this wave was not captured by the interaction effect of sex-state (see Fig. 9). States on the route of drug trafficking to the US like Sinaloa, Durango, Coahuila, Chihuahua, and Nayarit showed large state main effects (see Fig. 6).

Estimates of state-year interaction for Tamaulipas were high from 2010 to 2012, when massive burials were found in San Fernando Tamaulipas (see Fig. 10). This state has been dominated by the Zetas, a former military and extremely violent group. The states of Guerrero, Michoacán, and Oaxaca on the Pacific Coast also present high estimates of main state effects (see Fig. 6). This is consistent with the presence of the Beltran Leyva and La Familia cartels in that region [30]. These states are important producers of cannabis and poppy. Baja California also presents large state estimates. The city of Tijuana, where the Tijuana cartel is based, is located in the Baja California state. Tijuana is a very coveted place by other drug lords due to its strategic location.

Although female homicides do not reach the same levels as male homicides, causes cannot be associated entirely with the war against drugs. Chihuahua and State of Mexico had remarkable high rates of female homicide for younger groups of age, and may have a different cause. INEGI's data show that in most cases reported, homicide occurred in private spaces but are increasing in public spaces. This problem has been neglected by authorities.

Overall, since 2012 there is a slowdown in the trend (see Fig. 5). The government that started in 2012 and the one that started in 2018 also saw a continuous escalation of violence. However, the period from 2013 onwards is not considered in this analysis.

This study has revealed the population segments with the highest and lowest effects on homicides in different states and years. More research has to be done on the specific cause of death, i.e. type of homicide and place of occurrence (i.e. streets, alleys, main roads, highways, home, work, office, and factory). This can help to deepen knowledge on the nature of crime and contribute to addressing those vulnerable sectors with a gender perspective in order to prevent homicide and other types of violence. Also, aggregating into smaller geographical units like municipalities would reveal more knowledge about the dispersion of homicide. The National Program for Social Prevention of Violence and Delinquency was canceled in 2016, and the only municipality which implemented an integral social program to address violence and prevention of crime in 2010 was Ciudad Juárez. However, it consumed a huge amount of economic resources which made the program more difficult to implement and afford by other cities, municipalities, and states.

Finally, we would like to emphasize the importance of analyzing the statistical results with historical, geographical, and social events for a better understanding of homicide in Mexico and their effects on society.

References

1. Astorga, L.: Qué querían que hiciera? Inseguridad y delincuencia organizada en el gobierno de Felipe Calderón, 1st edn. Penguin Random House Grupo Editorial (2015)
2. Atuesta, L.H., Ponce, A.F.: Meet the Narco: increased competition among criminal organisations and the explosion of violence in Mexico. Glob. Crime **18**(4), 375–402 (2017). https://doi.org/10.1080/17440572.2017.1354520
3. Castillo, J.C., Mejia, D., Restrepo, P.: Scarcity without Leviathan: the violent effects of cocaine supply shortages in the Mexican drug war. Rev. Econ. Stat. **102**(2), 269–286 (2020). https://doi.org/10.1162/rest_a_00801
4. Celis, M.d.P.: Geografía de la violencia en México. un acercamiento a la reconfiguración territorial de la violencia generada por el crimen organizado (2016). https://politicadedrogas.org/documentos/20161106_164436_15_geografaaviolenciafinal_281016.pdf
5. Cressie, N.A.: Statistics for Spatial Data, revised Wiley, New York (1993)
6. Curiel, R.P., Delmar, S.C., Bishop, S.R.: Measuring the distribution of crime and its concentration. J. Quant. Criminol. **34**, 775–803 (2018). https://doi.org/10.1007/s10940-017-9354-9
7. Damián Bernal, A.L.: La manifestación espacial de la violencia feminicida en México, 2000–2006. Master thesis, UNAM, Facultad de Filosofía y Letras (2010)
8. Dell, M.: Trafficking network and the Mexican drug war. Am. Econ. Rev. **105**(6), 1738–79 (2015). https://doi.org/10.1257/aer.20121637
9. Durán-Martínez, A.: To kill and tell? State power, criminal competition and drug violence. J. Conflict. Resolut. 1–27 (2015). https://doi.org/10.1177/0022002715587047
10. Espinal-Enriquez, J., Larralde, H.: Analysis of Mexico's Narco-war network (2007–2011). PLoS ONE **10**(5), 1–15 (2015). https://doi.org/10.1371/journal.pone.0126503
11. Flores, M., Rodríguez-Oreggia, E.: Spillover effect on homicides across Mexican municipalities: a spatial regime model approach. Rev. Reg. Stud. **44**(3), 241–262 (2014). https://doi.org/10.52324/001c.8072
12. Fuerte-Celis, P., Sanchez-Castañeda, D.: Clusters of violence in Mexico: an analysis of homicide rates from 2000 to 2012. J. Lat. Am. Geogr. (2021). https://doi.org/10.1353/lag.2021.0004
13. García-Ponce, O., Dube, A., Dube, O.: Cross-border spillover: U.S. gun laws and violence in Mexico. Am. Polit. Sci. Rev. **107**(3), 397–417 (2013). https://doi.org/10.1017/S0003055413000178
14. Guerrero-Gutierrez, E.: Security, drugs, and violence in Mexico: a survey. In: 7th North American Forum. Washington, US (2011). https://lantiaintelligence.com/storage/document/15/NAF_2011.pdf
15. Ibrahim, J., Laud, P.: A predictive approach to the analysis of designed experiments. J. Am. Stat. Assoc. **89**(425), 309–319 (1994)
16. Ingram, M.C., Marchesini da Costa, M.: A spatial analysis of homicide across Brazil's municipalities. Homicide Stud. **21**(2), 87–110 (2017). https://doi.org/10.1177/1088767916666603
17. Lacombe, D.J., Flores, M.: A hierarchical SLX model application to violent crime in Mexico. Ann. Reg. Sci. **58**, 119–134 (2017). https://doi.org/10.1007/s00168-016-0788-z
18. Lawson, A.B.: Bayesian Disease Mapping: Hierarchical Modeling in Spatial Epidemiology. Chapman and Hall, CRC (2013). https://doi.org/10.1201/9781351271769

19. Leenen, I., Cervantes-Trejo, A.: Temporal and geographic trends in homicide and suicide rates in Mexico, from 1998 through 2012. Aggress. Violent Behav. **19**, 699–707 (2014). https://doi.org/10.1016/j.avb.2014.09.004
20. Liu, Y., Fullerton, T.M., Ashby, N.J.: Assessing the impacts of labor market and deterrence variables on crime rates in Mexico. Contemp. Econ. Policy **31**(4), 669–90 (2013). https://doi.org/10.1111/j.1465-7287.2012.00339.x
21. Lunn, D., Spiegelhalter, D., Thomas, A., Best, N.: The bugs project: evolution, critique and future directions (with discussion). Stat. Med. **28**, 3049–3082 (2009)
22. Márquez-Padilla, F., Pérez-Arce, F., Rodríguez-Castelán, C.: The (non-) effect of violence on education: evidence from the 'war on drugs' in Mexico (2015). https://doi.org/10.2139/ssrn.2793566
23. Montgomery, D.C.: Design and Analysis of Experiments, 4th edn. Wiley (1997)
24. Osorio, J.: The contagion of drug violence: spatiotemporal dynamics of the Mexican war on drugs. J. Conflict Resolut. **59**(8), 1403–32 (2014). https://doi.org/10.1177/0022002715587048
25. Pan, M., Widner, B., Enomoto, C.E.: Growth and crime in contiguous State of Mexico. Rev. Urban Reg. Dev. Stud. **24**(1–2), 21–64 (2012). https://doi.org/10.1111/j.1467-940X.2012.00188.x
26. Parker, R.N.: Poverty, subculture of violence, and type of homicide. Soc. Forces **67**(4), 983–1007 (1989). https://doi.org/10.1093/sf/67.4.983
27. Ríos Contreras, V.: How government structure encourages criminal violence: the causes of Mexico's drug war. Ph.D. thesis, Harvard University (2013). https://dash.harvard.edu/bitstream/handle/1/11156675/RiosContreras_gsas.harvard_0084L_10752.pdf?sequence=3&isAllowed=y
28. Snyder, R., Durán-Martínez, A.: Does illegality breed violence? Drug trafficking and state-sponsored protection rackets. Crime, Law Soc. Chang. **52**(3), 253–73 (2009). https://doi.org/10.1007/s10611-009-9195-z
29. Sturtz, S., Ligess, U., Gelman, A.: R2OpenBUGS a package for running OpenBUGS from R. J. Stat. Softw. **12**(3), 1–16 (2005). https://cran.r-project.org/web/packages/R2OpenBUGS/vignettes/R2OpenBUGS.pdf
30. Valdés, C.G.: Historia del narcotráfico en México, 3a. edn. Ed. Aguilar (2013)

Index

S. Nieto-Barajas (ed.), *Interdisciplinary Statistics in Mexico*, Springer Proceedings in Mathematics & Statistics 397, https://doi.org/10.1007/978-3-031-12778-6